Methods in Molecular Biology

Series Editor
John M. Walker
School of Life and Medical Sciences
University of Hertfordshire
Hatfield, Hertfordshire, UK

For further volumes:
http://www.springer.com/series/7651

For over 35 years, biological scientists have come to rely on the research protocols and methodologies in the critically acclaimed *Methods in Molecular Biology* series. The series was the first to introduce the step-by-step protocols approach that has become the standard in all biomedical protocol publishing. Each protocol is provided in readily-reproducible step-by-step fashion, opening with an introductory overview, a list of the materials and reagents needed to complete the experiment, and followed by a detailed procedure that is supported with a helpful notes section offering tips and tricks of the trade as well as troubleshooting advice. These hallmark features were introduced by series editor Dr. John Walker and constitute the key ingredient in each and every volume of the *Methods in Molecular Biology* series. Tested and trusted, comprehensive and reliable, all protocols from the series are indexed in PubMed.

Buckwheat

Methods and Protocols

Edited by

Alexander Betekhtin and Artur Pinski

Institute of Biology, Biotechnology and Environmental Protection, University of Silesia in Katowice, Katowice, Poland

Humana Press

Editors
Alexander Betekhtin
Institute of Biology, Biotechnology
and Environmental Protection
University of Silesia in Katowice
Katowice, Poland

Artur Pinski
Institute of Biology, Biotechnology
and Environmental Protection
University of Silesia in Katowice
Katowice, Poland

ISSN 1064-3745 ISSN 1940-6029 (electronic)
Methods in Molecular Biology
ISBN 978-1-0716-3793-7 ISBN 978-1-0716-3794-4 (eBook)
https://doi.org/10.1007/978-1-0716-3794-4

This Humana imprint is published by the registered company Springer Science+Business Media, LLC, part of Springer Nature.
The registered company address is: 1 New York Plaza, New York, NY 10004, U.S.A.

Paper in this product is recyclable.

Preface

Buckwheat belongs to the *Fagopyrum* genus and is a promising functional food source containing various phenolic compounds, such as rutin, quercetin, and C-glycosylflavones (orientin, isoorientin vitexin), which have a positive therapeutic or dietary effect for promoting human health. To date, 22 buckwheat species have been identified. Most of them are wild buckwheat species, and the two most widely cultivated species are *F. esculentum* (common buckwheat) and *F. tataricum* (Tartary buckwheat). Common buckwheat is more distributed than Tartary buckwheat, and its range extends from temperate Europe to Japan through the Indo-Myanmar region. In addition, *F. tataricum* is an excellent natural source of phenolic compounds, and, compared to *F. esculentum*, it has better nutritional value and antioxidant activities. It has been reported that *F. tataricum* contains many more flavonoid compounds than *F. esculentum*. In these methods chapters, we would like to describe the most recent methodology for buckwheat research.

Katowice, Poland

Alexander Betekhtin
Artur Pinski

Contents

Preface ..... *v*
Contributors ..... *ix*

1 Callus Induction Followed by Regeneration and Hairy Root Induction in Common Buckwheat 1
Denis O. Omelchenko, Elena S. Glagoleva, Anna Y. Stepanova, and Maria D. Logacheva

2 Immunostaining for Epigenetic Modifications in *Fagopyrum* Calli 15
Alicja Tomasiak, Katarzyna Sala, and Agnieszka Braszewska

3 Quantitative Analysis of Epigenetic Modifications in *Fagopyrum* Nuclei with Confocal Microscope, ImageJ, and R Studio 23
Alicja Tomasiak, Lea Sophie Berg, Katarzyna Sala, and Agnieszka Braszewska

4 Phenolic Determination in Proembryogenic Cell Complexes of Buckwheat Morphogenic Cell Culture with Osmium Tetroxide, Toluidine Blue O Dye, and Iron Chloride 35
Yulia A. Kostyukova

5 Protoplast Isolation, Culture, and Regeneration in Common and Tartary Buckwheat 45
Magdalena Zaranek, Reneé Pérez-Pérez, Joanna Malec, and Ewa Grzebelus

6 Immunohistochemical Detection of the Wall Components on the Example of Shoot Apical Meristem of *Fagopyrum esculentum* and *Fagopyrum tataricum* 57
Anna Milewska-Hendel, Ewa Kurczyńska, and Kamila Godel-Jędrychowska

7 Immunodetection of Cell Wall Components in Studies on Cell Wall Rebuilding in *Fagopyrum esculentum* and *Fagopyrum tataricum* 71
Anna Milewska-Hendel, Katarzyna Sala-Cholewa, and Reneé Pérez-Pérez

8 Visualization of *Fagopyrum esculentum* and *Fagopyrum tataricum* Chromosomes and Micronuclei 81
Reneé Pérez-Pérez and Jolanta Kwasniewska

9 Embryo Rescue Technique in Buckwheat *Fagopyrum esculentum* Moench 89
Galina Suvorova

10 Comprehensive Embryological Analyses of Common Buckwheat (*Fagopyrym esculentum* Moench) 97
Aneta Słomka, Klaudia Sychta, and Agnieszka Płażek

11 Phenol-Based Protein Extraction Method for Plant Proteomic Studies 107
Przemysław Kopeć, Monika Krzewska, and Agnieszka Płażek

12 Two-Dimensional Gel Electrophoresis in Studies of Flower and Leaf Proteome of Common Buckwheat 113
Przemysław Kopeć, Monika Krzewska, and Agnieszka Płażek

13 Measurement of the Kinetics of Chlorophyll *a* Fluorescence by an LED-Light Source Fluorimeter, Handy PEA 121
Marta Hornyák and Agnieszka Płażek

14 Measurements of Leaf Gas-Exchange Parameters Using Portable CIRAS-3 Infrared Gas Analyzer, with a Parkinson Leaf Chamber (PLC6) 127
Marta Hornyák, Maciej Grzesiak, and Agnieszka Płażek

15 Measurement of the Light Intensity and Spectrum Influence on Plant Growth and Secondary Metabolites of Common Buckwheat 133
Monika Kula-Maximenko, Marta Hornyák, and Agnieszka Płażek

Index *139*

Contributors

Lea Sophie Berg • *Institute of Plant Sciences, University of Bern, Bern, Switzerland*

Agnieszka Braszewska • *Institute of Biology, Biotechnology and Environmental Protection, Faculty of Natural Sciences, University of Silesia in Katowice, Katowice, Poland*

Elena S. Glagoleva • *Institute for Information Transmission Problems of the Russian Academy of Sciences, Moscow, Russia; Faculty of Biology, Lomonosov Moscow State University, Moscow, Russia*

Kamila Godel-Jędrychowska • *Institute of Biology, Biotechnology and Environmental Protection, Faculty of Natural Sciences, University of Silesia in Katowice, Katowice, Poland*

Ewa Grzebelus • *Department of Plant Biology and Biotechnology, Faculty of Biotechnology and Horticulture, University of Agriculture in Krakow, Krakow, Poland*

Maciej Grzesiak • *The Franciszek Górski Institute of Plant Physiology, Polish Academy of Sciences, Kraków, Poland*

Marta Hornyák • *W. Szafer Institute of Botany, Polish Academy of Sciences, Kraków, Poland*

Przemysław Kopeć • *The Franciszek Górski Institute of Plant Physiology, Polish Academy of Sciences, Kraków, Poland*

Yulia A. Kostyukova • *Kazan Institute of Biochemistry and Biophysics, FRC Kazan Scientific Center of RAS, Kazan, Russia*

Monika Krzewska • *The Franciszek Górski Institute of Plant Physiology, Polish Academy of Sciences, Kraków, Poland*

Monika Kula-Maximenko • *The Franciszek Górski Institute of Plant Physiology, Polish Academy of Sciences, Kraków, Poland*

Ewa Kurczyńska • *Institute of Biology, Biotechnology and Environmental Protection, Faculty of Natural Sciences, University of Silesia in Katowice, Katowice, Poland*

Jolanta Kwasniewska • *Plant Cytogenetics and Molecular Biology Group, Faculty of Natural Sciences, University of Silesia in Katowice, Katowice, Poland*

Maria D. Logacheva • *Institute for Information Transmission Problems of the Russian Academy of Sciences, Moscow, Russia; Center for Molecular and Cellular Biology, Skolkovo Institute of Science and Technology, Moscow, Russia*

Joanna Malec • *Department of Plant Cytology and Embryology, Faculty of Biology, Jagiellonian University in Kraków, Kraków, Poland*

Anna Milewska-Hendel • *Institute of Biology, Biotechnology and Environmental Protection, Faculty of Natural Sciences, University of Silesia in Katowice, Katowice, Poland*

Denis O. Omelchenko • *Institute for Information Transmission Problems of the Russian Academy of Sciences, Moscow, Russia*

Reneé Pérez-Pérez • *Institute of Biology, Biotechnology and Environmental Protection, Faculty of Natural Sciences, University of Silesia in Katowice, Katowice, Poland*

Agnieszka Płażek • *Department of Physiology, Breeding of Plant and Seed Science, University of Agriculture, Cracow, Poland*

Katarzyna Sala • *Institute of Biology, Biotechnology and Environmental Protection, Faculty of Natural Sciences, University of Silesia in Katowice, Katowice, Poland*

Katarzyna Sala-Cholewa • *Institute of Biology, Biotechnology and Environmental Protection, Faculty of Natural Sciences, University of Silesia in Katowice, Katowice, Poland*

ANETA SŁOMKA • *Department of Plant Cytology and Embryology, Institute of Botany, Faculty of Biology, Jagiellonian University in Kraków, Cracow, Poland*
ANNA Y. STEPANOVA • *Group of Specialized Root Metabolism, K.A. Timiryazev Institute of Plant Physiology RAS, Moscow, Russia*
GALINA SUVOROVA • *Federal State Budgetary Scientific Institution, Federal Scientific Center of Legumes and Groat Crops, Orel oblast, Russia*
KLAUDIA SYCHTA • *Department of Plant Cytology and Embryology, Institute of Botany, Faculty of Biology, Jagiellonian University in Kraków, Cracow, Poland*
ALICJA TOMASIAK • *Institute of Biology, Biotechnology and Environmental Protection, Faculty of Natural Sciences, University of Silesia in Katowice, Katowice, Poland*
MAGDALENA ZARANEK • *Institute of Biology, Biotechnology and Environmental Protection, Faculty of Natural Sciences, University of Silesia in Katowice, Katowice, Poland*

Chapter 1

Callus Induction Followed by Regeneration and Hairy Root Induction in Common Buckwheat

Denis O. Omelchenko, Elena S. Glagoleva, Anna Y. Stepanova, and Maria D. Logacheva

Abstract

This section describes a set of methods for callus induction followed by the successful regeneration of whole plants and obtaining a culture of transgenic hairy roots from buckwheat plants (*Fagopyrum esculentum* Moench.). Callus induction and regeneration are key steps for many biotechnological, genetic, and breeding approaches, such as genetic modification, production of biologically active compounds, and propagation of valuable germplasm. Induction of hairy roots using *Agrobacterium rhizogenes* is also an important tool for functional gene research and plant genome modification. While many efforts were invested into the development of the corresponding protocols, they are not equally efficient for different cultivars. Here, we have tested and optimized the protocols of callus induction, regeneration, and transformation using *A. rhizogenes* for a set of cultivars of *F. esculentum*, including wild ancestor of cultivated buckwheat *F. esculentum* ssp. ancestrale and a self-pollinated accession KK8. The optimal medium for callus induction is Murashige-Skoog basal medium with 3% sucrose which includes hormones 2,4-dichlorophenoxyacetic acid 2 mg/L and kinetin 2 mg/L; for shoot initiation 6-benzylaminopurine 2 mg/L, kinetin 0.2 mg/L, and indole-3-acetic acid 0.2 mg/L; for shoot multiplication 6-benzylaminopurine 3 mg/L and indole-3-acetic acid 0.2 mg/L; and for root initiation half-strength Murashige-Skoog medium with 1.5% sucrose and indole-3-butyric acid 1 mg/L. *A. rhizogenes* R1000 strain proved to be the most efficient in inducing hairy roots in buckwheat and T-DNA transfer from binary vectors. Seedling explants cut at the root area and immersed in agrobacterium suspension, as well as prickling the cotyledonary area with agrobacteria dipped syringe needle, are the most labor-effective methods of infection, allowing to initiate hairy root growth in 100% of explants.

Key words *Fagopyrum esculentum*, Agrobacterium-mediated transformation, Hairy roots, Callus, Plant regeneration

---

Denis O. Omelchenko and Elena S. Glagoleva contributed equally with all other contributors.

Alexander Betekhtin and Artur Pinski (eds.), *Buckwheat: Methods and Protocols*, Methods in Molecular Biology, vol. 2791,
https://doi.org/10.1007/978-1-0716-3794-4_1,

1 Introduction

F. esculentum, commonly known as buckwheat, is an important pseudocereal crop widely used as food and feed. It also produces diverse valuable secondary metabolites for pharmaceutical purposes. Given the importance of buckwheat, biotechnological methods, such as plant cell culture and regeneration and genetic modification, are extensively developing for this species (reviewed in [1]). However, the studies show that callus induction and regeneration efficiency greatly differs between cultivars (e.g., [2–4]), types of explants, and media used. This calls for further research that involves more cultivars and accessions of buckwheat. As for genetic modification, though several protocols have been proposed [5–8], they are far from being highly efficient and universally reproducible.

A promising method for obtaining stable non-chimeric transformants is the *Agrobacterium rhizogenes* transformation and further regeneration of transgenic plants from hairy roots. This approach was successfully used for potato [9], cotton [10], and dandelion [11]. Hairy roots can also be a handful instrument for testing the efficiency of genetic editing (e.g., [12]; for review, see [13]). As a first step to applying this approach on buckwheat, we present two complementary protocols: for the induction of callus and its subsequent regeneration and for the induction and selection of transgenic hairy roots carrying a marker GFP gene. The first protocol relies on the regeneration of viable plants out of callus obtained from cotyledon explants. It outlines all necessary steps, including seed sterilization, germination, induction of callus, stimulation of shoot and root morphogenesis, and acclimation of regenerated plants. As a second step, we provide a step-by-step methodology for the induction and maintenance of hairy root cultures from *F. esculentum* explants using *A. rhizogenes*. The protocol outlines the preparation and germination of buckwheat seeds, the preparation of explants, the transformation process, and the subsequent selection, maintenance of transformed hairy roots, and induction of transgenic callus. These protocols were tested on several accessions of common buckwheat, including self-pollinated accession KK8. For hairy root induction, two strains of *A. rhizogenes* were tested.

2 Materials

2.1 Plants and Bacteria

1. Ripened seeds of the *F. esculentum* plants. Utilize seeds that have been harvested within the past few years and stored under appropriate conditions (in a dry and dark ventilated environment at room temperature) until the time of the experiments.

2. Agrobacterium *A. rhizogenes* strain R1000 that carries a binary vector with a gene of interest and marker genes for the selection of bacteria (antibiotic resistance) and transformed plants (visual marker, e.g., GFP and/or herbicide/antibiotic resistance).

2.2 Equipment

1. Forceps with dull and sharp points 160–200 mm long, suitable for sterilization.
2. Scalpel or small scissors.
3. Autoclave.
4. Laminar flow cabinet with UV lamp.
5. Gas/alcohol burner or electric sterilizer.
6. Ninety-six percent ethanol for alcohol burner. Seventy percent ethanol to wipe the working area and tools.
7. Autoclavable bottles for culture media.
8. Plastic containers with ventilated lids 100–120 mm high, sterile and/or autoclavable.
9. Sterile 15 or 50 mL tubes.
10. Autoclavable 100–250 mL Erlenmeyer glass flasks with vented caps or cellulose/cotton stoppers.
11. Pipettes with adjustable volumes 2–20 μL, 20–200 μL, and 100–1000 μL.
12. Sterile pipette tips for the pipettes.
13. Syringes and filters 0.2 μm for sterilization of thermolabile components.
14. Manual or electronic pipette controllers and sterile serological pipettes of 2 mL, 10 mL, and 25 mL to transfer liquid culture media.
15. Shaking incubator.
16. Tube rocker/rotator.
17. Dry-air thermostat.
18. Climate chamber or growth room with the ability to maintain temperature and light regime of day and night.
19. Sterile Petri dishes 100 mm diameter.
20. Sterile filter paper.
21. Parafilm.

2.3 Reagents

1. Murashige-Skoog basal medium with vitamins and 3% sucrose (MS3): NH_4NO_3 1650 mg/L, KNO_3 1900 mg/L, $CaCl_2$ * $2H_2O$ 440 mg/L, $MgSO_4$ * $7H_2O$ 370 mg/L, KH_2PO_4 170 mg/L, KJ 0.83 mg/L L, H_3BO_3 6.2 mg/L, $MnSO_4$ * $5H_2O$ 24.1 mg/L, $ZnSO_4$ * $7H_2O$ 8.6 mg/L, Na_2MoO_4 *

$2H_2O$ 0.25 mg/L, $CuSO_4$ * $5H_2O$ 0.025 mg/L, $CoCl_2$ * $6H_2O$ 0.025 mg/L, $FeSO_4$ * $7H_2O$ 27.8 mg/L, Na_2EDTA * $2H_2O$ 37.26 mg/L, mesoinositol 100 mg/L, nicotinic acid (PP) 0.5 mg/L, pyridoxine-HCl (B6) 0.5 mg/L, thiamine-HCl (B1) 0.1 mg/L, glycine 2 mg/L, sucrose 30 g/L, MES 500 mg/L. Adjust the pH of the solution to 5.6–5.7 with 1 M KOH and sterilize by autoclaving. To prepare a solid medium after pH adjustment and before sterilization, add agar 7 g/L.

2. Half-strength Murashige-Skoog basal medium with vitamins and 1.5% sucrose (1/2MS1.5): contains the same components as MS3, but two times less. Adjust the pH of the solution to 5.6–5.7 with 1 M KOH, and sterilize by autoclaving. To prepare a solid medium after pH adjustment and before sterilization, add agar 7 g/L.
3. 1/2MS0: half-strength Murashige-Skoog basal medium with vitamins and without sucrose.
4. Agrobacterium infiltration medium (AIM): MS3 with 200 μM acetosyringone.
5. YEB medium: 5 g/L peptone, 1 g/L yeast extract, 5 g/L beef extract, 5 g/L sucrose, 0.49 g/L $MgSO_4 \cdot 7H_2O$. Could be substituted with LB medium: 10 g/L peptone, 5 g/L yeast extract, 10 g/L sodium chloride. To prepare solid medium, add agar 15 g/L.
6. 1 M KOH for pH adjustment and plant hormone preparation. Dissolve 5.6 g in 100 mL water to prepare a 1 M stock.
7. Sterile distilled water.
8. Acetosyringone $C_{10}H_{12}O_4$ 100 mM stock solution. Dissolve 0.1962 g in 10 mL DMSO.
9. Cefotaxime antibiotic solution 100 mg/mL. Dissolve in distilled water; sterilize by filtration through a 0.2 μm filter.
10. Basta (phosphinothricin) herbicide solution 10 mg/mL. In distilled water, sterilize by filtration through a 0.2 μm filter.
11. Fundazole fungicide solution 1 mg/mL with Tween-20. Add 50 mg of powder in 50 mL of sterile distilled water, and mix to a cloudy white suspension. Add 10 μL of Tween-20, and mix again. Use freshly prepared.
12. Lysoformin-3000 disinfectant solution 2%. Dilute 1 mL of concentrate in 50 mL of sterile distilled water. Use freshly prepared.
13. Plant growth regulators (PGR):
    (a) 2,4-Dichlorophenoxyacetic acid (2,4-D) 1 mg/mL stock
    (b) Kinetin 1 mg/mL stock.
    (c) 6-Benzyl amino purine (BAP) 1 mg/mL stock

(d) Indole-3-acetic acid (IAA) 1 mg/mL stock.

(e) Indole-3-butyric acid (IBA) 1 mg/mL stock.

To prepare PGR stock solution (1 mg/mL), dissolve 10 mg of PGR in 100 μL of 1 M KOH. Mix well until the whole sample is completely dissolved. Add distilled water to the final volume of 10 mL, and sterilize by filtration through a 0.2 μm filter (or autoclave with the medium). PGR stock solutions can be stored at 4 °C for ~1 month.

14. Callus induction medium (CIM): MS3 medium with 2 mg/L 2,4-D, 2 mg/L kinetin. To prepare a solid medium, add agar 7 g/L.
15. Shoot initiation medium (SIM): MS3 medium with 2 mg/L BAP, 0.2 mg/L kinetin, 0.2 mg/L IAA. To prepare a solid medium, add agar 7 g/L.
16. Shoot multiplication medium (SMM): MS3 medium with 3 mg/L BAP, 0.2 mg/L IAA. To prepare a solid medium, add agar 7 g/L.
17. Root initiation medium (RIM): 1/2MS1.5 medium with 1 mg/L IBA. To prepare a solid medium, add agar 7 g/L.

3 Methods

3.1 Seed Preparation, Germination, and Sprouting

3.1.1 Seed Preparation and Sterilization

1. Scrape off the remnants of the sepals from the seed, wash the buckwheat seeds in soapy water, and rinse with distilled water (Fig. 1a).
2. Place the seeds in a 50 mL tube, cover them with distilled water, and leave to soak for 2–3 h. It helps seeds to germinate successfully and makes it easier to dehull them (Fig. 1b).
3. Drain the water, and carefully dehull the seeds with the help of sharp pointed forceps, starting from the wide part of the seed, where there were remnants of the sepals, trying not to damage the seed inside. All manipulations with seeds after this step should be carried out under aseptic conditions in a laminar flow cabinet (Fig. 1c).
4. Place dehulled seeds in a sterile 50 mL conical tube, fill the tube with a fungicide solution (20–25 mL of Fundazole with Tween-20) to cover the seeds, and incubate in a tube rocker/rotator at 20 rpm for 20 min.
5. Drain the fungicide solution.
6. Fill the tube with a disinfectant solution (20–25 mL of 2% Lysoformin-3000) to cover the seeds, and incubate in a tube rocker/rotator at 20 rpm for 10 min.
7. Drain the disinfectant solution.

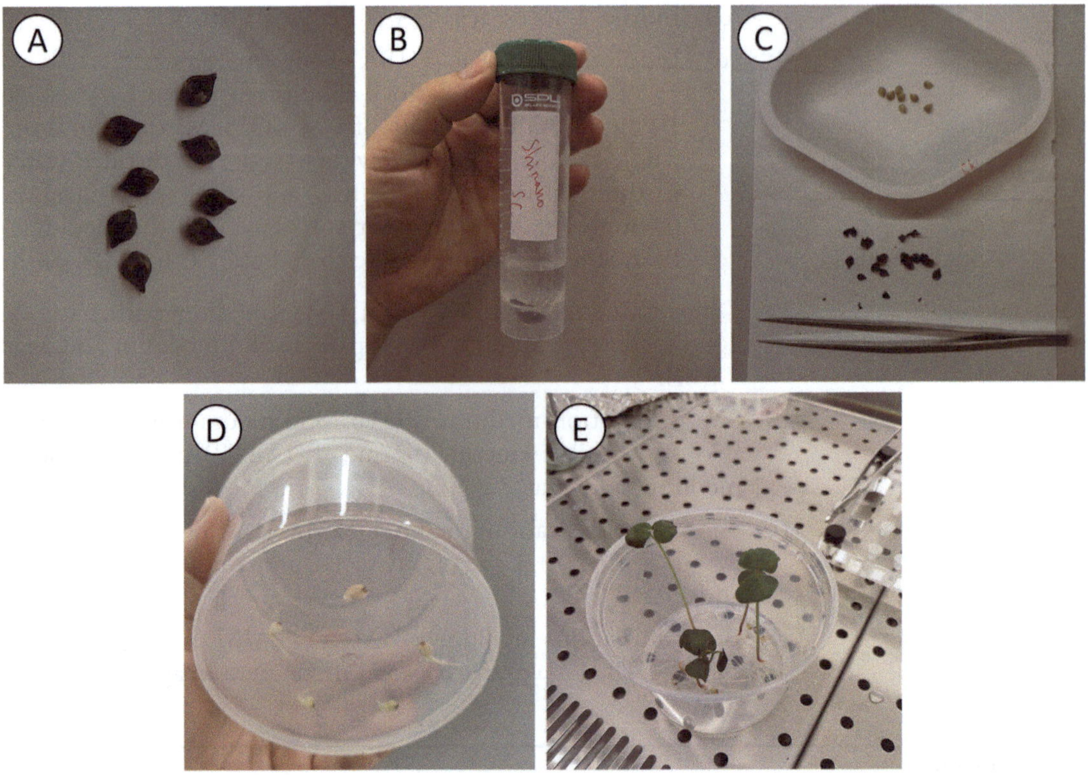

Fig. 1 Example of *F. esculentum* seed preparation and sprouting. *F. esculentum* Shinano seeds (**a**); soaking seed in water (**b**); dehulled imbibed seeds (**c**); sprouting seeds, 2 days after planting (**d**); seedlings, 5 days after planting (**e**)

8. Rinse the seeds with 30 mL of sterile distilled water four times to completely remove traces of the fungicide and disinfectant solutions.

3.1.2 Seed Sprouting

1. Using dull pointed forceps, place seeds evenly on a 1/2MS1.5 solid medium into sterile containers 100–120 mm high with a vented lid, up to six seeds per container.
2. Place containers with seeds in the dark for 2 days at +25 °C (Fig. 1d).
3. Move the containers to the climate chamber or growth room for 2–3 days at 16/8 h day/night light cycle (+25 °C, 15000 lux) until the formation of seedlings with open cotyledons (Fig. 1e).

3.2 Callus Induction and Regeneration (See Note 1)

3.2.1 Callus Induction (5–6 Weeks)

1. Take the seedlings out of the germination medium (1/2MS1.5), and place them in an empty Petri dish. Cut the cotyledons (or hypocotyls, or petioles) into pieces about 7–10 mm in size using a sterile scalpel blade and forceps. Avoid notable wilting of the explant tissue during the procedure.

2. Transfer the explants onto the solid CIM in the Petri dish. Roughly three to five explants are allowed per 60 mm Petri dish. Seal the Petri dishes using Parafilm. Incubate under dim light (600 lux) at 25 °C.
3. After the callus appearance, separate the calli from the remnants of the explant and transfer to a fresh CIM every 2–3 weeks (*see* **Note 2**, Fig. 2a).

Fig. 2 Example of callus induction and regeneration stages. Primary callus induction from cotyledon explants (**a**); secondary friable callus culture (**b**); first signs of morphogenesis (**c**–**d**); formation of various morphological structures in the morphogenic callus of *F. esculentum* on SIM (**e**–**h**); regenerant rooting (**i**); flowering in vitro (**j**); hardening in perlite (**k**); regenerant hardened to non-sterile soil conditions (**l**)

3.2.2 Shoot Initiation (1–2 Months)

1. Transfer small pieces of calli (approx. 5 mm) onto solid SIM in sterile containers with vented lids using sterile forceps. Incubate under light (16/8 h day/night cycle, 6000 lux) at 25 °C. Calli should be transferred to fresh media every 2 weeks. If the callus has significantly increased in size by the end of the subculturing cycle, it can be divided into smaller pieces before transferring them to a fresh medium (Fig. 2b).
2. After several subculturing cycles (depending on the plant cultivar and initial callus condition), structural heterogeneity and morphogenic structures will emerge (*see* **Note 3**, Fig. 2c–g).

3.2.3 Shoot Multiplication and Outgrowth (1–3 Months)

1. Transfer the calli with shoot initials onto solid SMM in sterile plant growth containers with vented lids. Propagules should be transferred to fresh media every 2–3 weeks (*see* **Note 4**, Fig. 2h).

3.2.4 Root Initiation (2–3 Weeks)

1. Isolate well-developed 3–5-cm-high regenerated shoots from callus residues. Transfer shoots to sterile containers with vented lids containing solid RIM. The basal ends of the shoots should be embedded into the medium. Incubate under light (16/8 h day/night cycle, 15,000 lux) at 25 °C (Fig. 2i, j).

3.2.5 Acclimation to Soil Conditions (1–2 Weeks)

1. When the roots have developed enough (~2 weeks), carefully remove the plants from the agar avoiding damage and drying of the roots.
2. Place plantlets into sterile plant growth containers with vented lids containing autoclaved perlite sand moistened with a 1/2MS0 without PGR. Incubate under light (16/8 h day/night cycle, 15,000 lux) at 25 °C for about 1 week (Fig. 2k).
3. After 1 week, take plants from perlite, wash roots gently with room temperature tap water, and transplant to pots with a coco substrate-vermiculite mixture (1:2) moistened with 1/2MS0 salt solution (lacking organics components) or any appropriate nutrient solution for hydroponics (*see* **Note 5**). Cover containers with transparent plastic bags or lids (*see* **Note 6**).
4. Remove the lid entirely once the plant has acclimated to the humidity levels within the climate chamber or growth room and the leaves no longer wilt uncovered (Fig. 2l).
5. Continue incubating the plants under standard conditions (25 °C with 16/8 h day/night cycle, 15,000 lux). Water regularly.

3.3 Hairy Root Induction

3.3.1 Transformation

1. Three days before transformation, streak *A. rhizogenes* R1000 strain (*see* **Note 7**) carrying the desired binary plasmid from glycerol stock, stored at −80 °C, on the surface of LB or YEB

Petri dishes containing appropriate antibiotics. Incubate for 2 days at 28 °C in a dry-air thermostat until individual colonies appear.

2. Using a sterile bacterial loop or pipette tip, transfer one colony to 10 mL liquid YEB medium containing appropriate antibiotics in a sterile 50 mL tube, and incubate overnight (~16 h) at +28 °C in a shaking incubator at 200 rpm.
3. Dilute the suspension 1:100 in YEB liquid medium with appropriate antibiotics and 200 μM acetosyringone in an Erlenmeyer flask with a vent cap, and incubate for 4–5 h at +28 °C on a shaking incubator at 200 rpm until OD600 ~ 0.5–0.6.
4. Centrifuge the bacteria for 20 min at 4000 g. Remove the supernatant and resuspend the bacteria pellet in an equal volume of AIM.
5. Pour agrobacterium suspension into 100-mm-diameter Petri dishes 25 mL per plate.
6. Using sterilized dull pointed forceps and a scalpel, cut the seedlings ~1 cm from the roots, and place the wound of the explant in the agrobacterium suspension for 30 min. To increase infection spread and amount of obtained hairy roots, inoculate explants around the cotyledonary node area using an insulin syringe needle dipped in agrobacterium suspension (Fig. 3a).
7. Remove excess bacterial suspension by blotting explants on a sterile filter paper.
8. Stick the explants into MS3 solid medium with 200 μM acetosyringone in sterile 100–120-mm-high culture containers with a vented lid.

3.3.2 Cocultivation

1. Incubate explants at room temperature in the dark for 2 days.
2. Immerse explants in sterile distilled water supplemented with antibiotic (500 μg/mL cefotaxime; *see* **Note 8**) in 50 mL conical tubes, and incubate on a tube rotator/rocker at 20 rpm for 20 min.
3. Drain the water; then using sterilized dull pointed forceps, remove the explants from the tubes; blot them on sterile filter paper; and stick evenly into MS3 solid medium with 500 μg/mL cefotaxime in sterile containers with a vented lids. Place in climate chamber or growth room at 16/8 h day/night light cycle (+25 °C, 6000 lux).
4. After 4–10 days, the first hairy roots should appear on the explants (Fig. 3b–c).

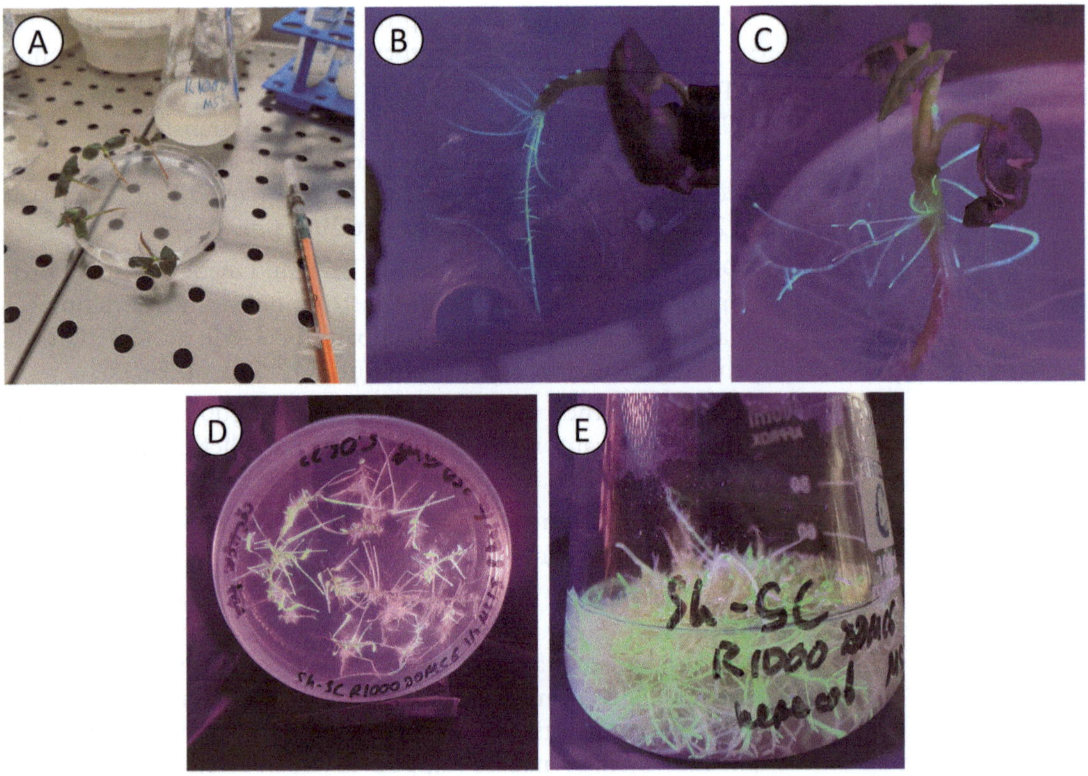

Fig. 3 Overview of hairy root induction protocol. Explant infection with *A. rhizogenes* agrobacterium suspension (**a**); hairy roots with and without transgene expression (GFP marker), 10 days after infection grown from root cut (**b**); hairy roots with and without transgene expression (GFP marker), 10 days after infection grown from needle prickled cotyledonary area (**c**); grown single hairy roots on a plate, 10 days after excision (**d**); mass culture of hairy roots grown in liquid medium (**e**)

3.3.3 Selection

1. Cut off single hairy roots ≥2 cm long from the explant expressing visible marker (e.g., GFP under UV light). In case of antibiotic/herbicide selection, place hairy roots on a selective medium – a Petri dish with a solid MS3 medium containing appropriate antibiotic/herbicide for transgene selection (*see* **Note 9**) and 250 μg/mL cefotaxime. Seal the Petri dishes using Parafilm and place in the dark at +25 °C (Fig. 3d).
2. After 1–2 weeks, cut the tip portion of grown hairy roots 2–3 cm long and repeat selection, because some of the hairy roots could be chimeric and escape selection. It is recommended to additionally check the expression of the gene of interest by qPCR and/or the presence of a visual marker (e.g., GFP).
3. Growing hairy roots can then be transplanted into an MS3 liquid medium and grown with slow shaking (90 rpm) on an incubator shaker at +25 °C in the dark to obtain a mass culture of hairy roots (Fig. 3e).

3.4 Callus Induction in Hairy Roots

Callus induction in hairy root culture could be achieved with different sets of PGR on MS3 medium, some of which are presented on Fig. 4.

1. Cut single root fragments from grown hairy root 1–3 cm long.
2. Place them on MS3 medium with appropriate set of PGR (Fig. 4).
3. After 2–3 weeks, first transgenic calli will start to appear on the surface of the hairy root fragments.
4. Transfer growing calli to fresh media every 2–3 weeks.

4 Notes

1. The duration for maintaining callus cultures on callus medium may vary depending on the purpose of use. Over time, the callus will acquire an increasingly friable and light texture, especially if grown in the dark. However, it's important to note that the morphogenic and organogenic capacity of the calli can diminish over time. Therefore, for successful plant regeneration, it is advisable to promptly transfer the calli to the shoot initiation medium. Usually, up to about five subculture cycles on a callus induction medium are recommended if one intends to use calluses for plant regeneration. However, in some instances, the morphogenicity of the callus can persist for extended periods.
2. The regeneration capacity of the different buckwheat cultivars may differ vastly. The protocol described here has been used to produce viable regenerated plants of several cultivars of *F. esculentum*, specifically Shinano, Batyr, and KK8. Some other cultivars of *F. esculentum* Moench. and *F. tataricum* Gaertn. have also been evaluated for their regeneration capacities, but the efficiency is lower. To optimize plant regeneration from a given cultivar, test different levels of plant growth regulators, BAP, in combination with IAA in shoot initiation medium.
3. Morphogenesis can be manifested by trichomes, shoot or root initials, leaf primordia, and embryonic structures (Fig. 2). Additionally, sections of the callus may exhibit reddening (anthocyanins synthesis) or greening. Further targeted isolation and cultivation of callus sites with these specific structures can increase the chances of plant regeneration. Buds with the potential to develop shoots can be identified without microscopy by their darker green color compared to the surrounding tissue.

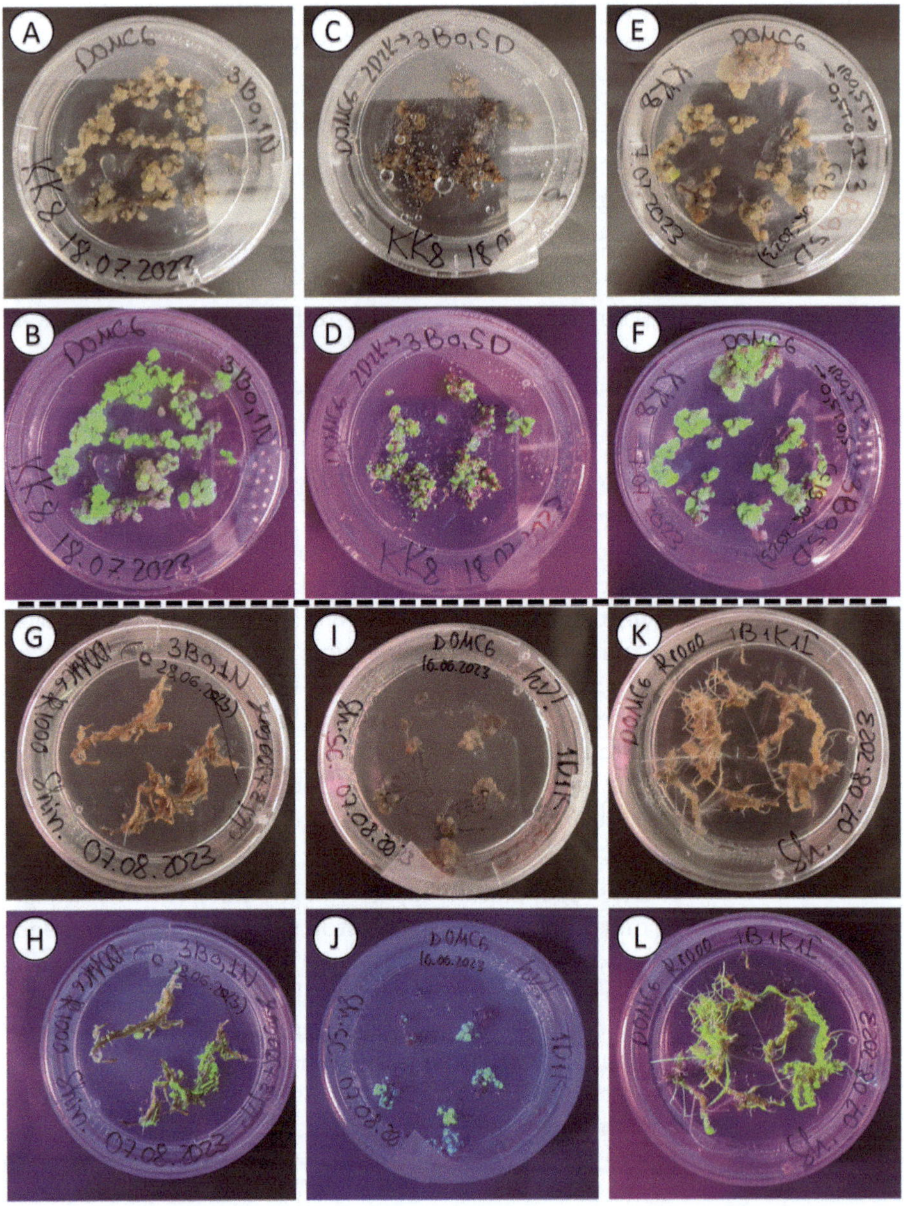

Fig. 4 Examples of transgenic calli grown from hairy root cultures of KK8 (**a–f**) and Shinano (**g–l**) cultivars. KK8 and Shinano cultured on MS3 medium with 3 mg/L 6-BAP and 0.1 mg/L NAA under daylight (**a**, **g**) and UV light (**b**, **h**); KK8 cultured in two different media in succession – MS3 2 mg/L 2,4-D and 2 mg/L kinetin, MS3 3 mg/L 6-BAP and 0.5 mg/L 2,4-D under daylight (**c**) and UV light (**d**); KK8 cultured in three different media in succession – MS3 1 mg/L 6-BAP and 5 mg/L thidiazuron, then MS3 0.5 mg/L thidiazuron and 0.5 mg/L IAA, then MS3 3 mg/L 6-BAP and 0.5 mg/L 2,4-D under daylight (**e**) and UV light (**f**); Shinano cultured on MS3 medium with 1 mg/L 2,4-D and 1 mg/L kinetin under daylight (**i**) and UV light (**j**); or Shinano cultured on MS3 medium with 1 mg/L 6-BAP, 1 mg/L kinetin, and 1 mg/L IAA under daylight (**k**) and UV light (**l**)

4. It is acceptable to multiplicate and outgrowth shoots on the same as shoot initiation medium or use Murashige & Skoog Shoot Multiplication Medium A (Sigma M7024).
5. To prevent plant fungal infection, a fungicide (e.g., Fludioxonil or other) can be added to the substrate.
6. To facilitate the hardening of the plants, it is recommended to periodically and briefly (for 1 or a few hours) remove the lid or plastic bag over 3 days following 3 days of growth. Alternatively, you can create small openings in the bag.
7. The A4 strain also successfully infects and causes the growth of hairy roots in the buckwheat; however, unlike R1000, no transgene expression (GFP) was observed with A4-induced hairy roots in any tested buckwheat plants in our experiments.
8. Cefotaxime can be substituted for Carbenicillin or Timentin or other antibiotics that have also been successfully used to inhibit agrobacteria growth.
9. We've found 4–6 mg/L of Basta optimal to prevent growth of the non-transgenic hairy roots, though it also reduces the rate of transgenic root growth. We recommend testing optimal Basta concentration for your specific cultivar.

Acknowledgments

This study was supported by the Ministry of Higher Education and Science of Russia (project # 075-15-2021-1064 and # FFNU-2022-0037). The authors are grateful to Ivan N. Fesenko and Alexey N. Fesenko for providing buckwheat seeds.

References

1. Tomasiak A, Zhou M, Betekhtin A (2022) Buckwheat in tissue culture research: current status and future perspectives. Int J Mol Sci 23: 2298
2. Srejović V, Nešković M (1981) Regeneration of plants from cotyledon fragments of buckwheat (*Fagopyrum esculentum* Moench). Zeitschrift für Pflanzenphysiologie 104:37–42
3. Gumerova EA, Galeeva EI, Chuyenkova SA et al (2003) Somatic embryogenesis and bud formation on cultured *Fagopyrum esculentum* hypocotyls. Russ J Plant Physiol 50:640–645
4. Woo SH, Nair A, Adachi T et al (2000) Plant regeneration from cotyledon tissues of common buckwheat (*Fagopyrum esculentum* Moench). In Vitro Cell Dev Biol Plant 36: 358–361
5. Kojima M, Arai Y, Iwase N et al (2000) Development of a simple and efficient method for transformation of buckwheat plants (*Fagopyrum esculentum*) using *Agrobacterium tumefaciens.* Biosci Biotechnol Biochem 64:845–847
6. Bratić AM, Majić DB, Miljuš-Đukić JD et al (2007) *In planta* transformation of buckwheat (*Fagopyrum esculentum* Moench.), vol 59, pp 135–138
7. Chen L-H, Zhang B, Xu Z-Q (2008) Salt tolerance conferred by overexpression of Arabidopsis vacuolar Na+/H+ antiporter gene *AtNHX1* in common buckwheat (Fagopyrum esculentum). Transgenic Res 17:121–132
8. Zhou B, Ding F, Yang Z et al (2023) Establishing a highly efficient *Agrobacterium*-mediated

transformation system in sweet buckwheat. J Consum Prot Food Saf

9. Butler NM, Jansky SH, Jiang J (2020) First-generation genome editing in potato using hairy root transformation. Plant Biotechnol J 18:2201–2209
10. Cui M-L, Liu C, Piao C-L et al (2020) A stable *Agrobacterium rhizogenes*-mediated transformation of cotton (*Gossypium hirsutum* L.) and plant regeneration from transformed hairy root via embryogenesis. Front Plant Sci 11:604255
11. Iaffaldano B, Zhang Y, Cornish K (2016) CRISPR/Cas9 genome editing of rubber producing dandelion Taraxacum kok-saghyz using *Agrobacterium rhizogenes* without selection. Ind Crops Prod 89:356–362
12. Zhang H, Cao Y, Zhang H et al (2020) Efficient generation of CRISPR/Cas9-mediated homozygous/biallelic *Medicago truncatula* mutants using a hairy root system. Front Plant Sci 11:294
13. Kiryushkin AS, Ilina EL, Guseva ED et al (2021) Hairy CRISPR: genome editing in plants using hairy root transformation. Plants (Basel) 11:51

Chapter 2

Immunostaining for Epigenetic Modifications in *Fagopyrum* Calli

Alicja Tomasiak, Katarzyna Sala, and Agnieszka Braszewska

Abstract

Immunostaining is a well-established technique for identifying specific proteins in tissue samples with specific antibodies to identify a single target protein. It is commonly used in research and provides information about cellular localization and protein expression levels. This chapter describes a detailed protocol for immunostaining fixed *Fagopyrum* calli embedded in Steedman's wax using nine antibodies raised against histone H3 and H4 methylation and acetylation on several lysines and DNA methylation.

Key words Immunostaining, Antibody, Callus, *Fagopyrum*, Histone methylation, Histone acetylation, DNA methylation

1 Introduction

Immunostaining is deployed in cell biology to research the differential expression of proteins and the localization and distribution at the subcellular, cellular (immunocytochemistry), and tissue (immunohistochemistry) levels [1, 2]. The principle of this technique used in our protocol employs the use of two types of antibodies. The primary antibody, isotype IgG, is targeted against a protein of interest to detect its presence by binding to it, whereas the secondary antibody, conjugated with a fluorescent tag, subsequently binds to the primary antibody, allowing it to be visualized or quantified using the fluorescence microscopy [3]. Primary antibodies that are typically developed in rabbits or mice have a high affinity for specific proteins and can bind to fixed tissues [4]. Secondary antibodies are usually raised in goats or donkeys and labeled with fluorophores such as Alexa Fluor dyes, enzymes such as horseradish peroxidase (HRP) or alkaline phosphatase, and proteins such as biotin to amplify the primary antibody's signal [3, 5–9]. Specimens are usually fixed in paraformaldehyde or methanol to prevent tissue

Alexander Betekhtin and Artur Pinski (eds.), *Buckwheat: Methods and Protocols*, Methods in Molecular Biology, vol. 2791, https://doi.org/10.1007/978-1-0716-3794-4_2,

degradation while maintaining and preserving the morphology and the epitopes they bind to. Samples can be mounted into different types of wax to enable slicing into microscopy sections [6, 10].

The immunostaining technique is well-established in plant research [6]. Here, using the secondary antibodies conjugated with Alexa 488 fluorochrome, we were able to measure the relative fluorescence intensity of the *Fagopyrum tataricum* morphogenic, non-morphogenic, and *Fagopyrum esculentum* embryogenic calli for seven histone modifications, DNA methylation, and two controls (unmodified histone H3 and H4).

2 Materials

2.1 Tissue Preparation

The tissues used in this protocol are morphogenic and non-morphogenic callus of *Fagopyrum tataricum* and embryogenic callus of *Fagopyrum esculentum* (Fig. 1).

1. Laminar flow chamber.
2. Laboratory stereoscopic Nikon microscope (Japan).
3. Forceps.
4. 2 mL Eppendorf tubes.

2.2 Tissue Fixation

1. 10× phosphate buffered saline (PBS): dissolve 80 g of NaCl, 2.0 g of KCl,14.4 g of Na_2HPO_4, 2.4 g of KH_2PO_4 in 800 mL distilled water. Adjust pH to 7.3 and the volume to 1 L with additional distilled H_2O. Sterilize by autoclaving, and keep at room temperature; for 1× PBS, use 10 mL of the 10× PBS with 900 mL of sterile H_2O.
2. 1× PBS – dissolve the following in 800 mL distilled H_2O: 8 g of NaCl, 0.2 g of KCl,1.44 g of Na_2HPO_4, 0.24 g of KH_2PO_4. Adjust pH to 7.3 and the volume to 1 L with additional

Fig. 1 *Fagopyrum tataricum* morphogenic callus (1**a**); white arrows: proembryogenic cell complex; black arrows: soft callus cells, non-morphogenic callus (1**b**), and *Fagopyrum esculentum* embryogenic callus (1**c**). Scale bar: 0.5 mm

distilled H_2O. Sterilize by autoclaving, and keep at room temperature (*see* **Note 1**).

3. 4% paraformaldehyde (PFA): dissolve 4 g of PFA in 100 mL of 1× PBS on the magnetic stirrer at 60 °C; adjust pH to 7.3 (*see* **Note 2**).
4. Vacuum pump.
5. 2 mL Eppendorf tubes.
6. Graded alcohol series: 10% EtOH in 1 × PBS, 30% EtOH in 1 × PBS, 50% EtOH in 1 × PBS, 70% EtOH in 1 × PBS, 90% EtOH in 1 × PBS, 99.9% EtOH.

2.3 Wax Embedding

1. Poly(ethylene glycol) distearate 400.
2. 1-Hexadecanol.
3. Laboratory incubator at 37 °C.
4. 99.9% EtOH.
5. Embedding moulds.
6. Heat plate.

2.4 Tissue Sectioning

1. HYRAX M40 rotary microtome.
7. Poly-L-lysine microscope slides.
8. Forceps.
9. Distilled H_2O.

2.5 Dewaxing

1. Graded alcohol series: 99.9% EtOH 70% EtOH in 1 × PBS, 50% EtOH in 1 × PBS, 30% EtOH in 1 × PBS.
2. 1× PBS.

2.6 Immunostaining

1. 5% bovine serum albumin (BSA): dissolve 2.5 g of BSA in 50 mL of 1 × PBS (*see* **Note 3**).
2. 1% BSA: dissolve 0.5 g of BSA in 50 mL of 1 × PBS (*see* **Note 4**).
3. Antibodies (*see* **Note 5**) (*see* Table 1).
4. 1:10 4′,6-diamidino-2-phenylindole (DAPI): dissolve 2.5 g of DAPI powder in 1 mL of 1× PBS (*see* **Note 6**).
5. Vectashield antifade buffer: mix 6 μL of 1:10 DAPI and 96 μL of Vectashield (*see* **Note 7**).
6. Foil/Parafilm cut to cover the slide.
7. Humid chamber.
8. 2N HCl.

Table 1
List of antibodies used in the immunostaining procedure

Antibody	Catalog number	Company
Anti-5-methylcytosine	ab73938	Abcam
Anti-dimethyl-histone H3 (lysine 4)	04–790	Sigma-Aldrich
Anti-trimethyl-histone H3 (lysine 4)	04–745	Sigma-Aldrich
Anti-trimethyl-histone H3 (lysine 36)	abe305	Sigma-Aldrich
Anti-acetyl-histone H3 (lysine 18)	ab1191	Abcam
Anti-acetyl-histone H4 (lysine 16)	ab109463	Abcam
Anti-acetyl-histone H4 (lysine 12)	04–119	Sigma-Aldrich
Anti-acetyl-histone H4 (lysine 5)	04–118	Sigma-Aldrich
Unmodified H3	ab18521	Abcam
Unmodified H4	ab10158	Abcam
Goat anti-mouse IgG	ab150113	Abcam
Goat anti-rabbit IgG	ab150077	Abcam

3 Methods

Carry out all procedures at room temperature unless otherwise specified.

3.1 Tissue Preparation

1. For morphogenic callus of *Fagopyrum tataricum*, pick the proembryogenic cell complexes (PECCs) carefully under the stereoscopic microscope with the use of fine forceps. For non-morphogenic callus of *Fagopyrum tataricum* and embryogenic callus of *Fagopyrum esculentum*, pick out bright pieces of about 2 mm.
2. Place the material in 2 mL Eppendorf tube (*see* **Note 8**).

3.2 Material Fixation

1. Fix the material in freshly prepared 4% paraformaldehyde (Sigma-Aldrich) in 1 × PBS, pH 7.3, and place in the vacuum desiccator for 3 h (*see* **Note 9**).
2. Carefully replace the fixative with 1 × PBS, and perform dehydration in a graded ethanol series diluted in 1 × PBS solution twice for 15 min in each concentration (10%, 30%, 50%, 70%, and 90%) and 99.8% twice for 30 min each.
3. Prepare the embedding medium: nine parts of poly(ethylene glycol) distearate 400 (Sigma-Aldrich) and one part of 1-hexadecanol (Sigma-Aldrich) (9/1 w/w) (*see* **Note 10**).

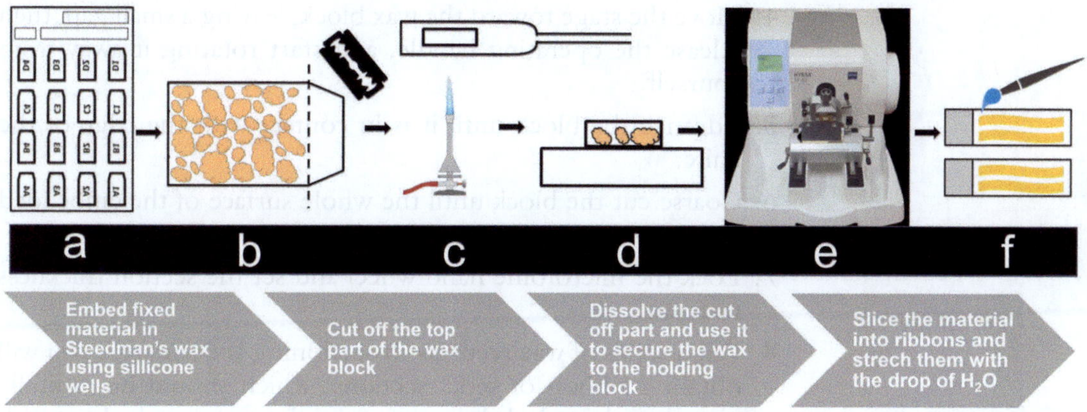

Fig. 2 General scheme of embedding and slicing the sample. (**a**) Silicone embedding well. Embedding should be performed on the heat plate set to 37 °C. (**b**) Place the material on the wider end of the well and leave the top empty. After the polymerization of wax overnight, take out the wax block and slice the top off. (**c**) Melt the top of the block over a flame; (**d**) pour the wax onto the wooden block and place the wax block on top. Leave at −20 °C for 5 min for the wax to solidify. (**e**) Secure the wooden block into the rotary microtome block holder, ensuring the wax block is at the right angle of tilt and in parallel with the blade. (**e**) Slice the block on the rotary microtome to the 5 μm ribbon of serial sections, (**f**) which should be carefully placed on the poly-L-lysine coated slides and stretched out with a drop of water

4. Perform the embedding procedure in the laboratory incubator set for 37 °C with a series of graded ethanol/wax – 3/1, 1/1, 1/3 v/v – for 24 h each, followed by two changes of pure wax for another 24 h.
5. For embedding of the material, place the embedding mould (Fig. 2a) on the heat plate set for 37 °C, wait for it to warm up, and subsequently place the material in the wider end of the mould (Fig. 2b) (*see* **Note 11**), and cover them with fresh wax so it is raised above the mould (*see* **Note 12**).
6. Leave the moulds to polymerize overnight at room temperature; store at 4 °C.

3.3 HYRAX M40 Rotary Microtome Slicing

1. Cut off the narrow end of the wax block (Fig. 2b) and dissolve it (Fig. 2c). Pour the liquid wax onto the wood block, and place the remaining wax block on top (side with the letter and number facing you) (Fig. 2d); put in −20 °C for 5 min.
2. Prepare the microtome, move the knife stage all the way to the back, place the blade in the knife holder and secure it by pulling a lever, and set the settings to trim at 10 μm (Fig. 2e).
3. Secure the wooden block in the block holder and ensure the wax block is at the right angle of tilt and in parallel with the blade.

4. Move the stage toward the wax block, leaving a small gap, then release the operating handle, and start rotating it away from yourself.
5. Advance the block until it is in contact with the edge of the knife.
6. Coarse cut the block until the whole surface of the embedded tissue can be sliced.
7. Lock the microtome hand wheel and set the section thickness to 5 μm.
8. Cut a series of wax sections. If sectioning is done well, you will obtain a ribbon of serial sections, which should be carefully placed on the poly-L-lysine coated slides and stretched out with several drops of water, tilting it gently from one side to another (Fig. 2f) and left to dry overnight.
9. Continue until the block is sliced; label the slides with a pencil, avoiding pens with nonalcohol-resistant ink. Slides should be stored at 4 °C.

3.4 De-embedding and Immunostaining

1. Place the slides in 99.8% ethanol three times for 10 min (30 min in total), followed by rehydration in ethanol/ 1 × PBS solutions: 90%, 70%, 50%, 30% v/v and 1 × PBS for 10 min each.
2. After de-embedding for **DNA methylation**, perform an additional incubation with 2N HCl for 45 min at room temperature (*see* **Note 13**).
3. Incubate the slides with 5% BSA in 1 × PBS for 1 h in the humid chamber, vortex BSA before use, and cover the slides with foil to avoid evaporation.
4. Apply 100 μL of the primary antibody in 1% BSA in 1 × PBS (1: 100), and incubate at 4 °C overnight in the humid chamber.
5. The next day, wash the slides 3 × 5 min in 1 × PBS, and apply 100 μL of the secondary antibody in 1% BSA in 1 × PBS (1: 100).
6. Incubate the samples at 37 °C in the humid chamber in the dark for 1 h.
7. After the incubation, wash the slides 3 × 5 min in 1 × PBS, gently dry them, and close the slides with 40 μL of 1:10 DAPI (2.5 g/mL) in Vectashield antifade buffer using a 24 × 50 mm cover slide, very slowly to avoid trapping any air. Store at 4 °C in the dark.

4 Notes

1. 10 × PBS can be prepared and used as the stock solution for 1 × PBS; store at room temperature; however, we recommend preparing 1 × PBS since 10 × PBS can precipitate with time.
2. It might be necessary to add several microcrystals of NaOH to dissolve the PFA, and ensure to adjust the pH afterward; PFA dissolves in water, releasing formaldehyde in the process; preparation under a fume hood is required.
3. Prepared 5% BSA can be stored in 2 mL Eppendorf tubes at −20 °C.
4. Prepared 1% BSA can be stored in 2 mL Eppendorf tubes at −20 °C.
5. Antibodies should be divided into 5 μL aliquots and stored at −20 °C. This prevents detrimental to the antibody thawing and refreezing procedures.
6. This should be performed in the dim light. The solution can be stored at −20 °C.
7. This should be performed in the dim light; prepare fresh each time.
8. The ratio of the material to the fixative should be 1:20 for the correct fixation of the material.
9. The tube should remain open, turn off the vacuum pump after each hour, and release the pressure. The material should go down the tube.
10. To melt the ingredients, use the heat plate set to 50–55 °C, and stir them gently as they melt.
11. In the case of callus, the positioning is not significant; for other types of tissue, make sure the positioning is done meticulously at the right angle.
12. Each well is labeled and subsequently engraved in the wax block to help identify.
13. This should be done under a fume hood.

Acknowledgments

This work was supported by the National Science Centre Poland, project OPUS19 grant number 2020/37/B/NZ9/01499.

References

1. Choe B-K, Cho Y (1994) Immunostaining cells and tissues. In: Antibody Techniques, pp 259–272
2. Maity B, Sheff D, Fisher RA (2013) Immunostaining: detection of signaling protein location in tissues, cells and subcellular compartments. Methods Cell Biol 113:81–105
3. Cooper M, Lummas S (2017) Antibodies for immunochemistry. In: Renshaw S (ed) Immunohistochemistry: essential methods. Wiley, Cambridge, pp 1–24
4. Burry RA (2010) Immunocytochemistry: a practical guide for biomedical research. Springer, New York
5. Braszewska-Zalewska A, Bernas T, Maluszynska J (2010) Epigenetic chromatin modifications in brassica genomes. Genome 53:203
6. Braszewska-Zalewska AJ, Wolny EA, Smialek L, Hasterok R (2013) Tissue-specific epigenetic modifications in root apical meristem cells of *Hordeum vulgare*. PLoS One 8: e69204
7. Houben A, Belyaev ND, Turner BM, Schubert I (1996) Differential immunostaining of plant chromosomes by antibodies recognizing acetylated histone H4 variants. Chromosom Res 4: 191
8. Oliver C, Jamur MC (2010) Overview of antibodies for immunochemistry. Methods Mol Biol (Clifton, N.J.) 588:27–32
9. Vaughn K (2013) Immunocytochemical techniques. In: immunocytochemistry of plant cells, pp 1–41
10. Steedman HF (1957) Polyester wax; a new ribboning embedding medium for histology. Nature 4574:1345

Chapter 3

Quantitative Analysis of Epigenetic Modifications in *Fagopyrum* Nuclei with Confocal Microscope, ImageJ, and R Studio

Alicja Tomasiak, Lea Sophie Berg, Katarzyna Sala, and Agnieszka Braszewska

Abstract

Epigenetic programming plays a vital role in regulating pluripotency genes, which become activated or inactivated during the processes of dedifferentiation and differentiation during an organism's development. The analysis of epigenetic modifications has become possible through the technique of immunostaining, where specific antibodies allow the identification of a single target protein. This chapter describes a detailed protocol for the analysis of the epigenetic modifications with the use of confocal microscopy, subsequent image, and statistical analysis on the example of *Fagopyrum* calli with the use of nine antibodies raised against histone H3 and H4 methylation and acetylation on several lysines as well as DNA methylation.

Key words Confocal microscopy, Epigenetics, R Studio

1 Introduction

Epigenetic modifications, such as DNA methylation and histone modifications, influence the accessibility of the DNA and change the chromatin structure, regulating the gene expression as a result. These processes are indispensable for properly developing and differentiating distinct cell types in an organism. In the previous chapter, we described the detailed protocol for immunostaining the *Fagopyrum* calli. Using this protocol, we obtained images and performed a statistical analysis of the fluorescence intensity [1]. Subsequently, we performed statistical analysis representing the dynamic global dynamic fluctuations in levels of DNA methylation and histone modifications in the calli of *Fagopyrum* with different capacities for morphogenesis. The advantage of the confocal laser scanning microscope is obtaining high-resolution images of the optical sections through fluorescently labeled nuclei

Alexander Betekhtin and Artur Pinski (eds.), *Buckwheat: Methods and Protocols*, Methods in Molecular Biology, vol. 2791, https://doi.org/10.1007/978-1-0716-3794-4_3,

(z-stacks) [2, 3]. The fluorescent tags are excited by the laser beam, causing them to emit light, which is then screened through the pinhole aperture. Other light produced by the sample outside the focal plane is rejected with the screening sheet. A dichromic mirror then perceives the photons which pass the pinhole aperture in the microscope chamber, which directs it to a photomultiplier tube and next to the computer. The image is generated pixel-by-pixel and is of exceptional quality [3]. It allows to visualize the fluorescence distribution matching the particular histone modifications with respect to three-dimensional chromatin distribution in the architecture of a nucleus. Confocal microscope computer software is a tool which allows to view and process microscopic images. It provides features for image acquisition, visualization, and analysis for quantitative measurement of the fluorescent intensity. Acquired images must undergo further processing. This was done using Fiji/ImageJ, a Java-based image processing program established at the National Institutes of Health and the Laboratory for Optical and Computational Instrumentation (LOCI, University of Wisconsin) [4–7]. It is equipped with a comprehensible interface designed to perform many complex tasks such as editing, processing, and analyzing 8-bit color and grayscale, 16-bit integer, and 32-bit floating point images in various formats. ImageJ is designed to calculate area and pixel value statistics of selections and intensity-thresholded objects defined by the user. It contains standard image processing functions such as convolution, edge detection, contrast manipulation, sharpening, smoothing, logical and arithmetical operations, and median filtering [8–10]. Raw data obtained from the images can also be analyzed using different statistical analysis software. Here, we used R Studio, an integrated development environment (IDE) for R, a programming language for statistical analysis with its graphical representation developed by RStudio PBC [11]. R is mainly known as the statistics language, which is why R dominates over other programming languages in the development of statistical tools. It contains a variety of packages suitable for scientific data analysis, such as agricolae (Statistical Procedures for Agricultural Research), which can be used to find significant differences between means [12] and other packages dedicated to data visualization [13]. It allows for the easily reproducible process of obtaining data summary after the statistical analysis and data visualization in the form of graphs [14].

Implementing this protocol, we analyzed the epigenetic makeup of *Fagopyrum esculentum* embryogenic callus as well as *Fagopyrum tataricum* morphogenic and non-morphogenic calli during the course of the passage at four time points for histone and DNA methylation and histone acetylation.

2 Materials

1. Confocal microscope Olympus FV1000 (Olympus, Poland) confocal system equipped with an Olympus IX81 inverted microscope, 60× PlanApo oil immersion objective lens (NA 1.35), a 50 mW, 405 nm diode laser (excitation 405 nm), a 100-mW multiline (excitations 457/488/515 nm) argon ion laser (Melles Griot BV, The Netherlands) and two separate photomultipliers (R6357, Hamamatsu, Japan).
2. Olympus Fluoview ™ FV10-ASW 4.2 software, version 04.02.03.06.
3. ImageJ version 1.53 s software (Wayne Rasband, National Institutes of Health, USA).
4. R Studio 2022.12.0 Build 353.

3 Methods

3.1 Image Acquisition

The parameters should always be identical to obtain the automated segmentation of nuclei based on threshold values (a border value of pixels fluorescence intensity between the background and the object). Using the provided image acquisition software, set the following parameters:

1. Scan mode: unidirectional XYZ scan.
2. LUT set to Hi-Lo to ensure the lack of oversaturation. Now, the image is displayed using this specific look-up table where red pixels represent intensity beyond scale (saturated) and blue pixels represent 0 pixel value.
3. Adjust the levels of the detector sensitivity (HV) such that the image shows no red pixels.
4. Aspect ratio 1:1, scan speed 12.5 us/pixel, pixel resolution 512 × 512, zoom 1.0, rotation 0, time scan interval: FreeRun; sequential scan mode: Frame; it scans sequentially frame-by-frame and finishes scanning one channel first and then scans the next one, filter mode: Analog Int wide aperture which enhances the resolution and contrast of microscopy by using a confocal pinhole to eliminate out-of-focus light (*see* **Notes 1** and **2**).
5. For excitation of Alexa 488 fluorochrome select the 488 nm laser, set to 8%. For excitation of DAPI, use 405 nm laser, set to 4% from each channel device adjustment.
6. Gain set to 1× and offset to 1% on both channels.
7. Step size set to 0.3.

3.2 Image Analysis

The fluorescence intensities of Alexa 488 were measured as the mean values from Integrated Density (IntDen) with the threshold value parameter per one nucleus, which depicted the sum of all the pixels within the region of interest.

1. Import the image into the ImageJ: Select File→ Import→ Image sequence (Fig. 1a).
2. In the new window that appears (Fig. 1b), select the directory of the files; select 16-BIT in Type and Filters s_C001 for DAPI and s_C002 for Alexa488 (Fig. 1c).
3. To prepare the image from the stacks obtained with the confocal microscope, Z Project changed the projection type to Max intensity, and the image type should be 8-bit. In the ImageJ panel, select Image→ Stacks→Z project→ Max intensity (Fig. 1d and e).
4. Again, in the panel, select Image→ 8-bit (Fig. 1f).
5. Such prepared image should be saved in the TIFF format, which will be needed in the later stages of this protocol. Save the image: File→ Save as...→ TIFF.
6. Subsequently, prepare the image for the analysis – invert the image: Image→ Invert (Fig. 1g) (*see* **Note 3**).
7. Adjust the threshold: Image→Adjust Threshold (use default Black&White) →Apply (leave this window open) (Fig. 1h) (*see* **Note 4**).
8. If necessary, in the case of several nuclei merged, apply the watershed separation, a technique for cutting apart connected components into separate ones (Fig. 1i).
9. To measure the fluorescence intensity, set the following parameters: Analyze→ Set Measurements → Integrated density (the most significant parameter), Perimeter, Median, Area, Mean Gray Value (Fig. 1j) (*see* **Note 5**).
10. To analyze the particles, set the following parameters: Analyze→ Analyze particles→ set size to 5–50,000, circularity 0.2–1.0, select Add to manager, show overlay→ Display results (Fig. 1k) (*see* **Note 6**).
11. You will have two new windows appear: Results and ROI Manager. These Results are not the final ones; thus, the Results window and the inverted image window should be closed, and the changes NOT saved (Fig. 1l). The window marked in the red rectangle will be used in the next step; therefore, do not close it.
12. Open your MAX_intensity image again (not inverted) by dragging it to the ImageJ panel (Fig. 1m).
13. In the ROI window, unselect Show all on the bottom.

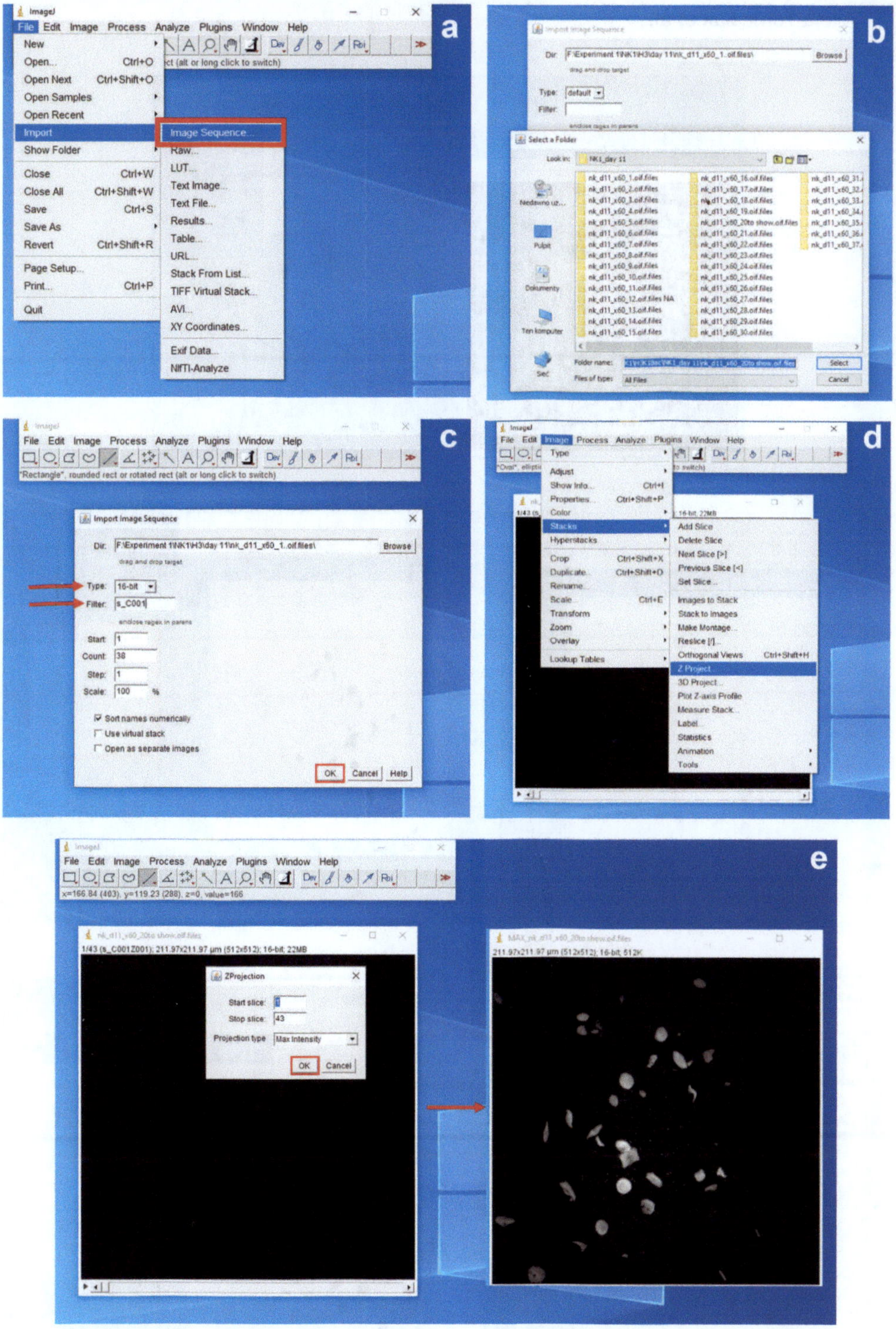

Fig. 1 (**a**) Uploading and saving the image; (**b**) selecting an image from the directory; (**c**) image type selection, filter for channels DAPI and Alexa488 application; (**d**) to prepare the image from the stacks, Z Project Max intensity should be selected, and the image type should be changed to 8-bit; (**e**) window appearing after Z Project. . . is selected, projection type Max intensity; (**f**) changing the image type to 8-bit; (**g**) image inversion; (**h**) threshold set up; (**i**) watershed application; (**j**) setting the measurements and the parameter selection; (**k**) setting parameters for the particle analysis; (**l**) windows appearing after particle analysis; (**m**) dragging previously saved .tiff image to the ImageJ panel, verification of the measured particles; (**n**) R Studio interface, Installing the packages from the Repository (CRAN); (**o**) results of the statistical analysis, graphs generated for the analysis

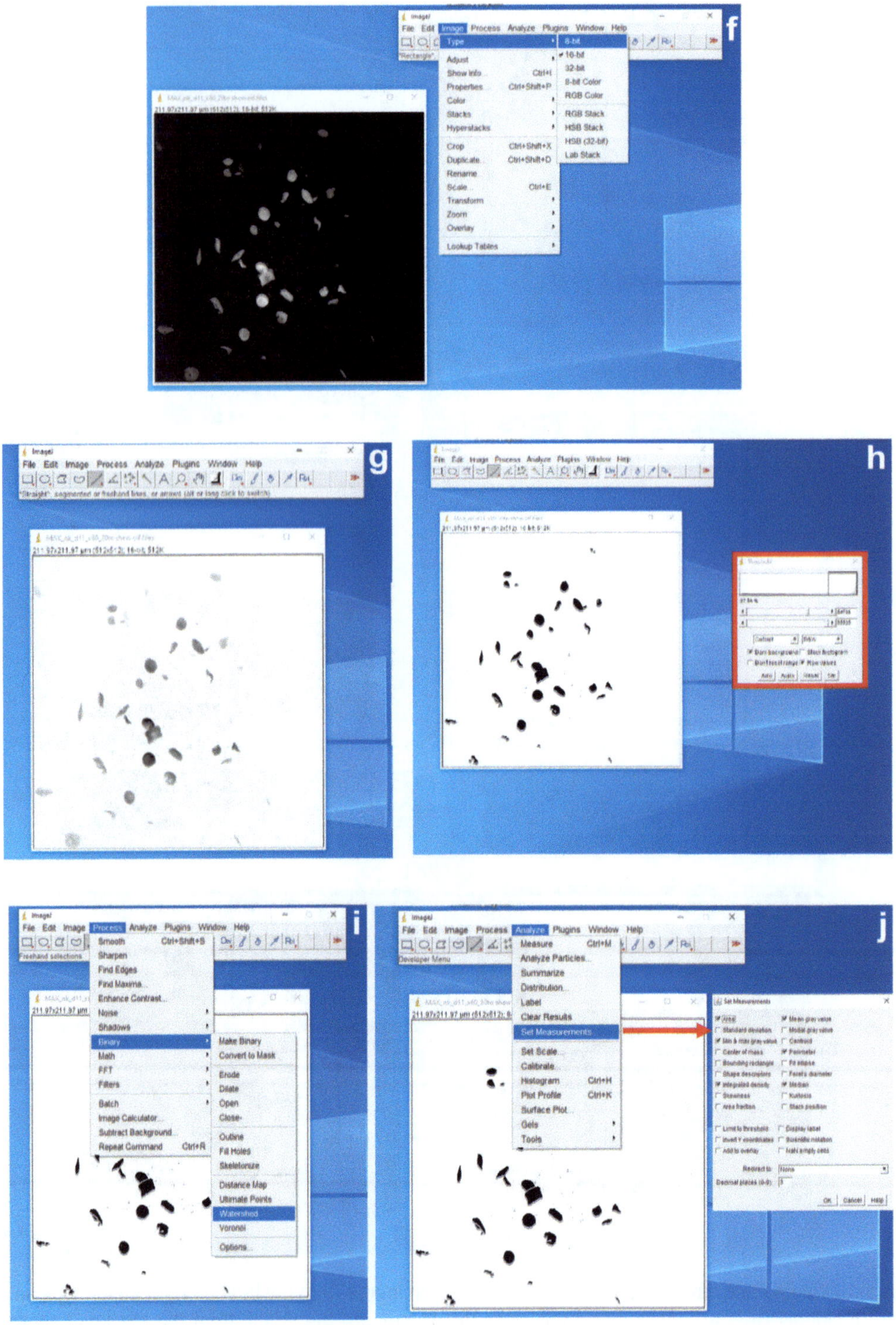

Fig. 1 (continued)

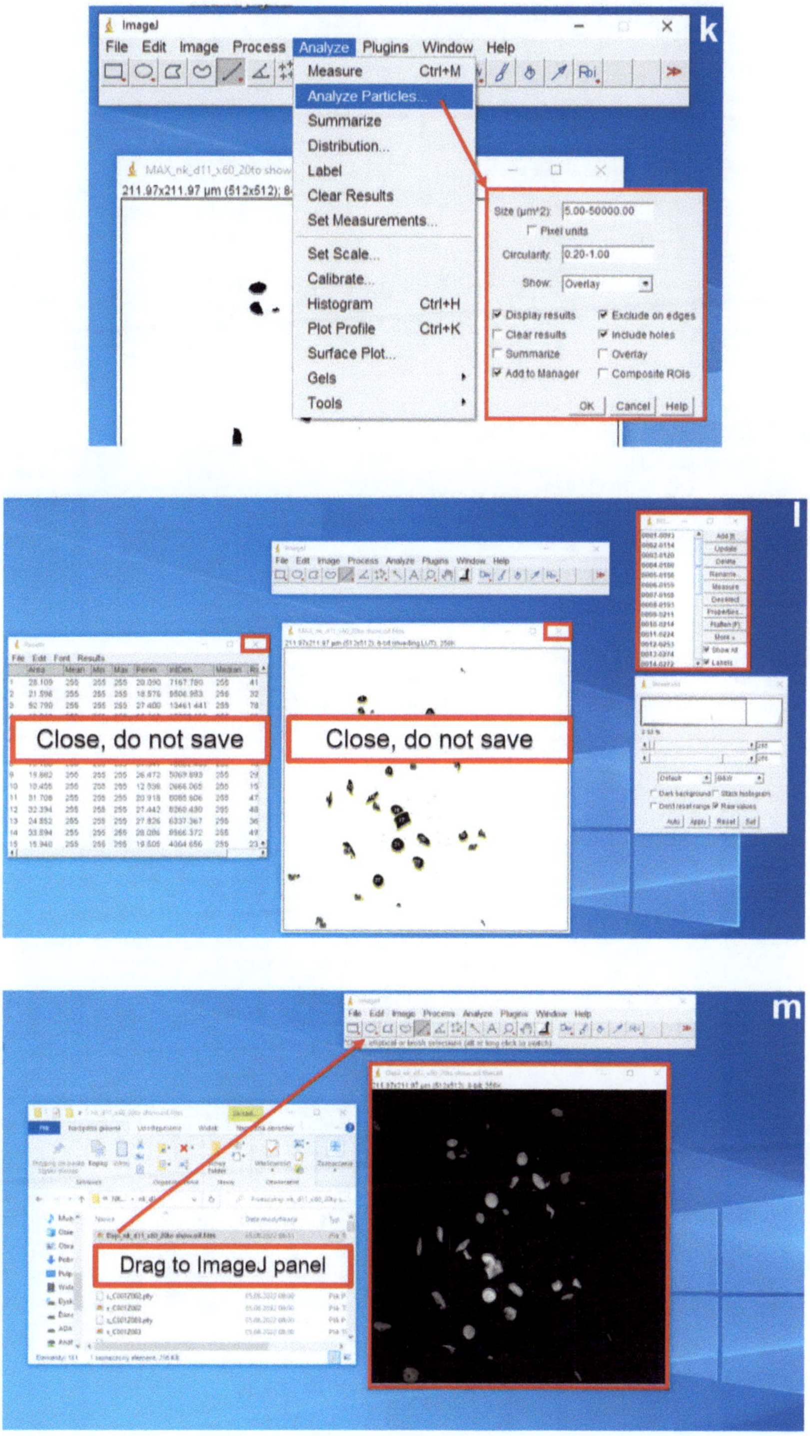

Fig. 1 (continued)

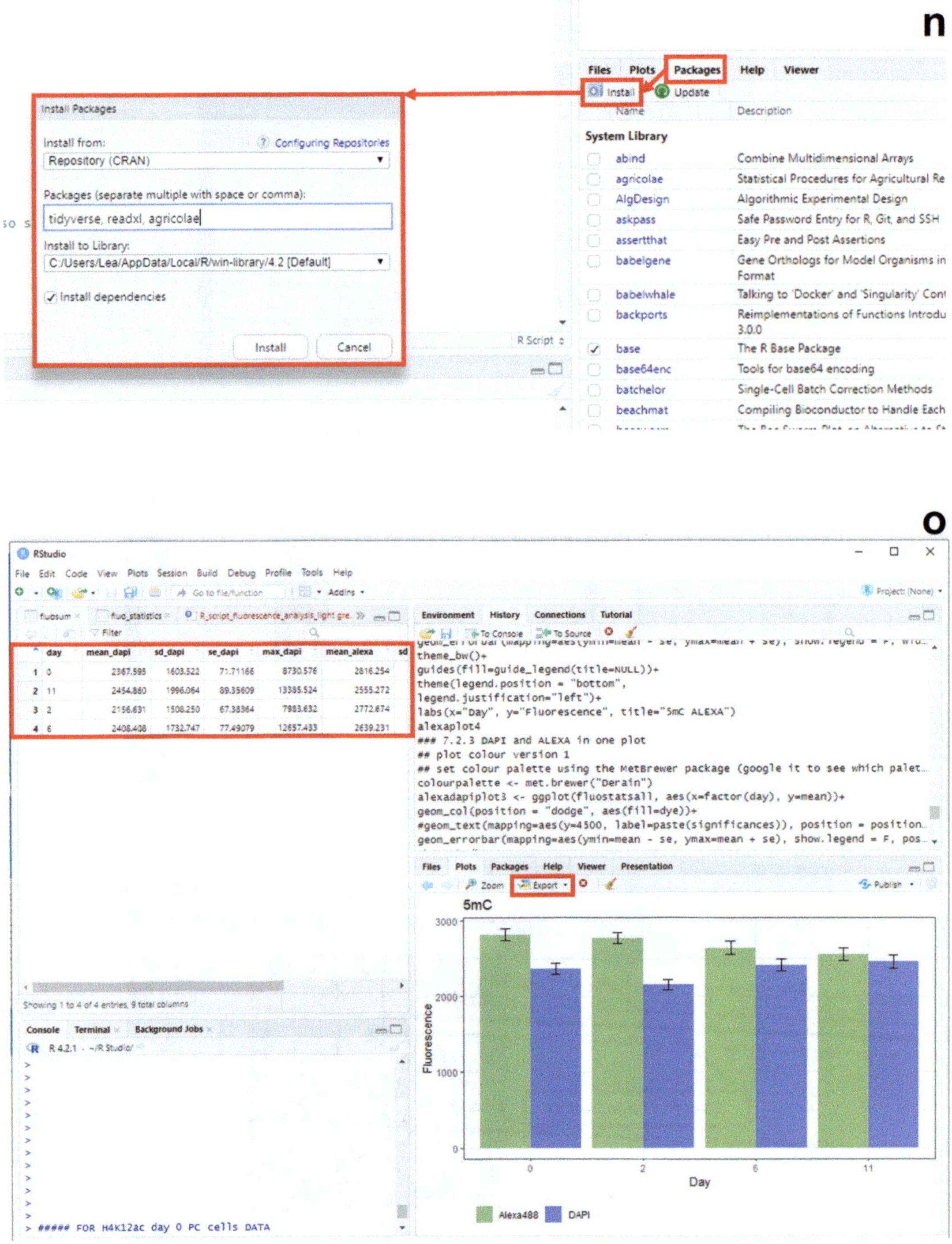

Fig. 1 (continued)

14. Go through each item on the list and delete the ones that are not nuclei, partial nuclei, nuclei that are merged.
15. Once this step is finished, select all remaining measurements on the list and press Measure; the new table will pop up; go to File→Save as.... Once the file is opened, the column IntDen will contain your measurements.
16. Repeat **steps 10–22** for the second filter (Alexa488, s_C002).

3.3 Statistical Analysis and Visualization with R

1. Open R Studio, and start a new R script by clicking the icon in the "File" category (*see* **Note 7**) in which you will write the required code. Click on "Packages" in the taskbar in the bottom right and download the required packages: tidyverse, readxl, and agricolae (Fig. 1n).
2. Load the libraries for these packages by using "library(packagename)" (*see* **Note 8**).
3. Load the Excel file with the data by using "dataset <- read_-excel("filename.xlsx", header = T)." The Excel file should contain a column that groups the nuclei, for example, the column "day," a column with the DAPI fluorescence values (name can be "dapi") and another column with the Alexa488 values ("alexa").
4. Make sure the columns have the correct class by using "as.numeric(dataset$dapi)" and "as.numeric(dataset$alexa)." The grouping column should be "as.character(dataset$day)."
5. To calculate mean fluorescence with standard deviation, execute the following code: "dataset % > % group_by(day) % > %

   summarise(mean_dapi = mean(na.omit(dapi)),

   sd_dapi = sd(na.omit(dapi)),

   mean_alexa = mean(na.omit(alexa)),

   sd_alexa = sd(na.omit(alexa)))" (*see* **Note 9**).
6. Next, use ANOVA and Tukey's HSD test to test for significant differences between the groups, in this case between days:

   "anovadapi <-aov(dataset$dapi ~dataset$day)

   summary(anovadapi)

   tukeydapi <- HSD.test(anovadapi, trt="dataset$day")

   tukeydapi"
7. The output shows each group (day) with a letter. There is a significant difference ($p < 0.05$) between groups when they have different letters (e.g., one group is "a" and the other is "b").
8. Repeat steps 44 and 45, and replace "dapi" in the code with "alexa" to get the test results for Alexa488 fluorescence.

9. Save the significance levels in a dataset like in the following example:

   "significances <- data.frame(day = c("0", "2", "6", "11"),

   significance_dapi = c("a", "b", "a", "a"),

   significance_alexa = c("a", "a", "b", "c"))"

10. Merge this with the original dataset:

    "final_data <- merge(dataset, significances, by = "day")"

11. Reorder the data to display the days in chronological order in the future plot:

    "final_data$day <- ordered(final_data$day, levels=c("0", "2", "6", "11"))"

12. To plot the DAPI data as bar plots with error bars, use the following code:

    "ggplot(final_data, aes(x = factor(day), y = mean_dapi))+

    geom_col()+

    geom_errorbar(mapping = aes(ymin = mean_dapi - sd_dapi, ymax = mean_dapi + sd_dapi), show.legend = F, width = 0.2)+

    labs(x ="Day", y="Fluorescence", title="DAPI fluorescence")"

13. Replace "dapi" in the previous code with "alexa" to show the Alexa488 results.

14. Executing this block of code will display the statistical analysis results and the plot in the "Plots" panel of R Studio where you can click on "Export" in order to save the plot as a file or to the clipboard (Fig. 1o).

4 Notes

1. For the quantitative analysis, the lower resolutions ensure faster scans and are sufficient for such analysis. For qualitative analysis, it is good to increase the resolution.

2. If you already have acquired images previously, it is possible to use the same parameters by right-clicking on one file from your previous acquisition and selecting "Load acquisition parameters" from the bottom of the list.

3. This can also be done by using Ctrl+Shift+I.

4. This can also be done by using Ctrl+Shift+T.

5. Once the settings are applied, this step can be skipped. Ensure the parameters are set each time the ImageJ is launched.

6. Once the settings are applied, this step can be skipped. Ensure the parameters are set each time the ImageJ is launched.
7. This can also be done by using Ctrl+Shift+N.
8. The packages must be loaded each time R Studio launches.
9. To display standard error instead of standard deviation, replace "sd_dapi = sd(na.omit(dapi))" with "se_dapi = sd(na.omit (dapi))/sqrt(length(na.omit(dapi))" and the same for alexa in the above code.

Acknowledgments

This work was supported by the National Science Centre Poland, project OPUS19 grant number 2020/37/B/NZ9/01499.

References

1. Braszewska-Zalewska A, Bernas T, Maluszynska J (2010) Epigenetic chromatin modifications in Brassica genomes. Genome 53:203
2. Pawley JB (2006) Handbook of biological confocal microscopy, 3rd edn. Springer, New York
3. Wilson T (2011) Resolution and optical sectioning in the confocal microscope. J Microsc 244:113
4. Abràmoff MD, Magalhães PJ, Ram SJ (2004) Image processing with imageJ. Biophoton Int 11:36–42
5. Collins TJ (2007) ImageJ for microscopy. BioTechniques 43:S25
6. Bourne R (2010) ImageJ. In: Fundamentals of digital imaging in medicine. Springer, London
7. Bankhead P (2014) Analyzing fluorescence microscopy images with ImageJ. ImageJ
8. Ferreira T, Rasband W (2012) ImageJ user guide user guide ImageJ. ImageJ User Guide 1:46r
9. Hartig SM (2013) Basic image analysis and manipulation in ImageJ. In: Current protocols in molecular biology
10. Schroeder AB, Dobson ETA, Rueden CT, Tomancak P, Jug F, Eliceiri KW (2021) The ImageJ ecosystem: open-source software for image visualization, processing, and analysis. Protein Sci 30:234
11. Martin G (2021) R Studio. In: An introduction to programming with R
12. Mendiburu F (2021) Statistical procedures for agricultural research. Wiley, New York
13. Kassambara A. (2022) Visualization of a correlation matrix using "ggplot2." Package 'ggcorrplot'
14. Del TI, Dickson K, Hakes AS, Newman SL (2022) Early undergraduate biostatistics & data science introduction using R, R Studio & the Tidyverse American Biology Teacher. Am Biol Teach 84:124–129

Chapter 4

Phenolic Determination in Proembryogenic Cell Complexes of Buckwheat Morphogenic Cell Culture with Osmium Tetroxide, Toluidine Blue O Dye, and Iron Chloride

Yulia A. Kostyukova

Abstract

The study of the localization of secondary metabolites in both plants and the cell cultures on the intravital sections is hampered by the difficulty of obtaining thin, correctly oriented sections. Techniques for fixing tissues in resins allow these difficulties to be overcome. Properly selected tissue fixation techniques allow using different dyes to identify the compound of interest. In addition, some components of tissue fixation can act as fixatives and as a dye for identifying secondary metabolites. For example, osmium tetroxide, which fixes lipids in tissues, stains phenolic compounds black. This paper describes methods for the detection of phenolic compounds in morphogenic callus culture of buckwheat using osmium tetroxide, Toluidine Blue O dye, and ferric chloride as dyes in epoxy resin-embedded cell culture with double fixation of the material and when material fixed in Karnovsky's fixative.

Key words Histochemical methods, Semi-thin sections, Phenolic, Osmium tetroxide, Toluidine Blue O dye, Ferric chloride

1 Introduction

Fagopyrum tataricum (L.) Gaertn. is a plant with intensive phenolic metabolism. Almost all classes of flavonoids as well as phenolic acids are synthesized in this plant [1, 2]. Phenolic acids have a wide range of therapeutic properties [2, 3]. Knowledge of the location of phenolic in plant tissues opens the possibility for more effective application of buckwheat plant for therapeutic purposes and also provides information for basic research on the synthesis and accumulation of phenolic. Callus and suspension cultures are suitable subjects for studying phenolic metabolism features because the cell culture's metabolism is similar to that of the parent plant [4]. In addition, maintaining cell cultures in vitro opens up the possibility of obtaining plant material regardless of the season.

Alexander Betekhtin and Artur Pinski (eds.), *Buckwheat: Methods and Protocols*, Methods in Molecular Biology, vol. 2791,
https://doi.org/10.1007/978-1-0716-3794-4_4,

The morphogenic callus of *Fagopyrum tataricum* (L.) Gaertn comprises proembryogenic cell complexes (PECCs) and a "soft" callus. The surface of PECCs is covered by cells that accumulate phenolics in their vacuoles [5]. The identification of phenolic compounds can be achieved by the use of different methods of tissue fixation and different dyes.

Phenolics are clearly visualized on histological semi-thin sections by double fixation of the material – in glutaric aldehyde and OsO_4. OsO_4 forms stable chelate complexes with o-dihydroxyphenols, which are visualized as electron-dense complexes on histological and electron-microscopic sections [6]. Using Karnovsky's fixative opens up the possibility of using Toluidine Blue O day and ferric chloride to identify phenolic compounds. Toluidine Blue O dye stain lignified cell walls (e.g., tracheids) in green and phenolic in green-blue color [7, 8]. Phenolic can also be identified by prefixing tissues in a 3% $FeCl_3$ solution [9]. $FeCl_3$ forms greenish-brown complexes with phenolics [10].

In this publication, we present a method for the identification of phenolic in buckwheat embryogenic callus cultures by prefixing with OsO_4, Toluidine Blue O day staining of callus tissues fixed in Karnovsky's fixative, and identification of phenolic by prefixing of callus tissues of buckwheat in ferric chloride fixed in Karnovsky's fixative. The proposed techniques were applied to plant tissue culture embedded in Epon-812 epoxy resin. Epon-812 epoxy resin has a relatively low viscosity and easily penetrates plant tissues [11, 12]. Because of these properties, Epon-812 epoxy resin is a widely used application for routine histological work [13].

The paper presents some of the subtleties of material fixation.

2 Materials

2.1 Plant Material

The PECC of morphogenic cell culture of buckwheat *F. tataricum* (L.) Gaertn.

2.2 Materials

1. 0.1 M Na – phosphate buffer (PhB) pH 7.2. The sodium phosphate buffer consists of two components: a monobasic sodium phosphate (NaH_2PO_4) and a dibasic sodium phosphate (Na_2HPO_4). Prepare stock solutions of the salts: NaH_2PO_4, 2.758 g/100 mL H_2O, and Na_2HPO_4, 2.838 g/ 100 mL H_2O. To prepare the buffer, add 28 mL of NaH_2PO_4 and the 72 mL of Na_2HPO. The stock solution of each salt adjusts the pH 7.2. After the volume of the buffer was adjusted to 200 mL with distilled water.
2. 0.2 M Na – phosphate buffer (PhB) pH 7.2. Prepare stock solutions of the salts: NaH_2PO_4, 2.758 g/100 mL H_2O, and

Na_2HPO_4, 2, 838 g/100 mL H_2O. Add 28 mL of NaH2PO4 and 72 mL of Na2HPO to prepare the buffer. The pH 7.2 is adjusted by the stock solution of each salt.

3. Glutaraldehyde (GA): 2.5% GA prepared on 0.1 PhB pH 7.2; 1 mL of 25% GA + 9 mL of PhB gives 10 mL of 2.5% GA.
4. Osmium tetroxide (OsO_4): a stock solution of 1% OsO_4 is prepared by dissolving 1 gram of crystalline osmium tetroxide in 100 mL of 0.1 M PhB pH 7.2.
5. Ethanol: dehydrating solutions are prepared at 20%, 30%, 40%, 50%, 60%, 70%, and 95% concentrations. Solutions are stored in the refrigerator and used cold during the dehydration process.
6. Acetone.
7. Epoxy resin 812: Epoxy resin 812 consists of epoxy resin (originally called Epon 812), the hardeners dodecenylsuccinic anhydride (DDSA) and methyl nadic anhydride (MNA), and an accelerator such as benzyldimethylamine (DMP – 30). Prepare Epoxy 812 resin embedding medium by pouring measured amounts into a graduated, polypropylene tube such as a 50 mL tube. A mixture of Epoxy resin 812 consists of the following components: Epoxy resin 812 20 mL, DDSA 7 mL, MNA 13 mL, and DMP-30 0.7 mL. The mixed components are stirred at room temperature for 1–2 h on a mixer (*see* **Note 1**).
8. Propylene oxide.
9. Toluidine Blue O dye [14]: prepare the 0.5% toluidine blue solution in 0.1% sodium carbonate (pH 11.1). Dissolve the sodium carbonate in the water, add the staining powder, and stir until dissolved. Filter before use.

   −0.5% toluidine blue is prepared by dissolving 500 mg of toluidine blue in 100 mL 0.1% sodium carbonate.

   −0.1% sodium carbonate is prepared by dissolving 100 mg of sodium carbonate in 100 ml distilled water.
10. Karnovsky's fixative [15]: mix the paraformaldehyde with 25 mL distilled water in a 125 mL Erlenmeyer flask. Heat to 60 °C on a hotplate. When moisture forms on the sides of the flask, add the sodium hydroxide and stir until the solution is clear. Cool the solution filter and then add GA and 0.2 M Na-phosphate buffer pH 7.2.
11. The 3% $FeCl_3$: the 3% $FeCl_3$ is prepared by dissolving 3 g of $FeCl_3$ in 100 mL of distilled water.

3 Methods

3.1 Fixation Material (See Note 2)

(a) Place the piece of plant tissue in embedding capsules with GA for 3 h in refrigerator cold (4 °C). If necessary, the tissues can be stored in the refrigerator for 1–2 weeks in the GA.

(b) Wash piece of plant tissue in embedding capsules with 0.1 M sodium phosphate buffer, pH 7.2, for 1 h. Gently pipette the GA from embedding capsules, and add the 0.1 M sodium phosphate buffer, pH 7.2.

3.2 Postfixation

After washing, the tissues are postfixed with 1% osmium tetroxide in 0.1 M sodium phosphate buffer for 3 h at room temperature. The 25 mg/mL of sucrose is added in 1% osmium tetroxide (*see* **Notes 3** and **4**).

3.3 Dehydration

1. Carefully remove the fixative solution with pipet from the embedding capsules, and dehydrate the plant tissue in an ethanol series. The ethanol dehydration solutions are prepared in increasing concentrations:

   20% – for 10 min

   30% – two changes, 10 min each

   50% – two changes, 15 min each

   60% – two changes, 15 min each

   70% – two changes, 15 min each

   95% – four changes, 15 min each.

2. Pipet the ethanol from the embedding capsule, and add the acetone three times, 20 min each.
3. Pipet the acetone from the embedding capsule, add propylene oxide, and incubate for 1 h at room temperature.

3.4 Embedding and Polymerization

Tissues are initially infiltrated with a combination of acetone and embedding medium. The infiltration can be carried out according to the following scheme:

2:1 – a mixture of acetone and embedding medium for 24 h in a refrigerator

1:1 – a mixture of acetone and embedding medium for 24 h in a refrigerator

1:2 – a mixture of acetone and embedding medium for 24 h in a refrigerator.

Monitor the formation of air bubbles in the capsule; remove them with a wooden stick (*see* **Note 5**).

Add the new mixture after completely removing the previous mixture. Place the plant tissue in a full embedding medium for polymerization at a temperature of 37 °C for 24 h, then 45 °C for 24 h, and 60 °C until sufficient hardness is reached.

The semi-thin sections from the hardened tissue blocks are prepared by sectioning with ultramicrotomes and are collected on microscope slides.

3.5 Staining of Section with Toluidine Blue O Dye

Place the section of plant tissue on the slide in a drop of Toluidine Blue O dye solution for 1–2 min at 60 °C. Rinse with distilled water.

3.5.1 Identification of Phenol in PECC by Toluidine Blue O Dye with Prefixation in Karnovsky's Fixative (See Note 6)

1. Transfer the piece of cell culture into Karnovsky's fixative in 1 h. If necessary, tissues can be stored in the refrigerator for 1–2 weeks.
2. For dehydration, embedding, and polymerization in Epoxy 12 resin, follow **steps 3.3–3.4** by the method described above**.**

3.5.2 Identification of Phenol in PECC Prefixed with $FeCl_3$ and Fixed in Karnovsky's Fixative [16] (See Note 7)

1. The PECC of the morphogenic callus of buckwheat is fixed in a mixture of 3% $FeCl_3$ for 1 h. The samples are washed in 0.1 M PhB pH 7.2.
2. The samples are placed in Karnovsky's fixative for 1 h.
3. After washing, the piece of plant culture is dehydrated and embedded in Epoxy −12 resin by method described above (see Subheadings 3.3 and 3.4).

4 Notes

1. The Epon812, DDSA, and MNA in the required volume are placed in a dry flask with a ground cover, stirred with a stirrer until a uniform color is formed (10 min), and then DMP-30 is added. All components are stirred for 2 h.
2. The phenolic in PECC can be identified on the semi-thin epoxy sections by double fixation of the tissue – in glutaraldehyde and osmium tetroxide. On semi-thin sections, the phenolic and osmium complexes appear as osmiophilic inclusions in cell vacuoles. For a clearer picture, histological sections can be stained with Toluidine Blue day (Fig. 1).
3. In case of the prefixation with 1% OsO_4, the solution's osmolarity is adjusted with sucrose for 400 mosm (25 mg of sucrose is added in 1 mL solution of 1% OsO_4).
4. For this reason, perfusion fixation with GA and secondary fixation with osmictetroxide are often performed with refrigerated cold (4 °C) solutions.
5. It is necessary to mix carefully with a wooden stick, avoiding the formation of air bubbles, in order to thoroughly

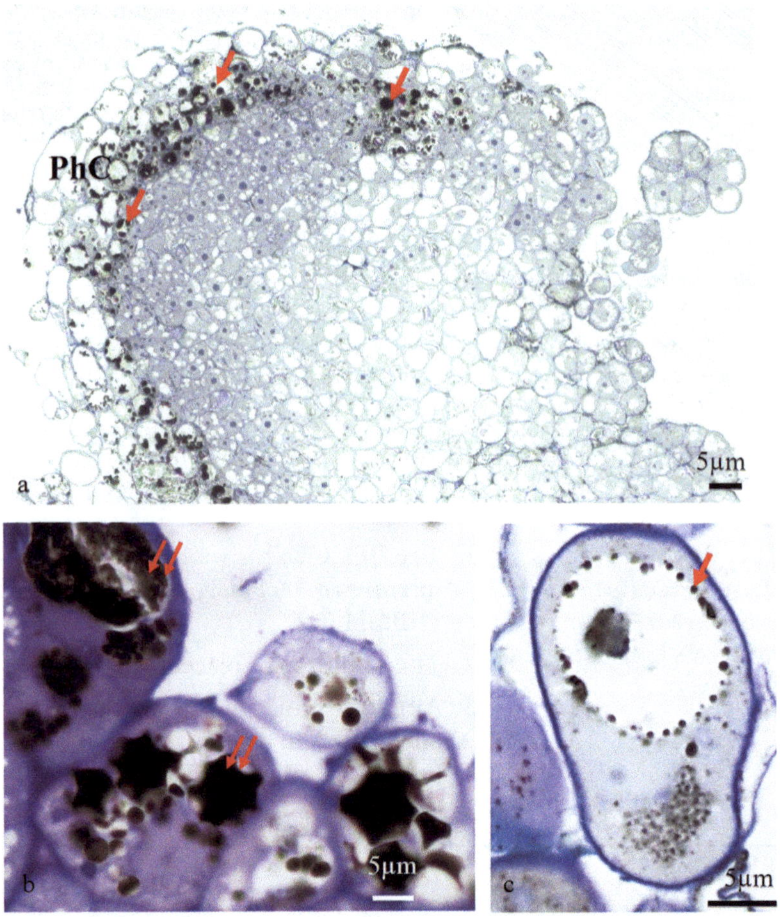

Fig. 1 The phenolic identification in proembryogenic cell complexes (PECCs) of morphogenic cell culture of buckwheat *Fagopyrum tataricum* (L.) Gaertn prefixed with OsO_4; (**a**) the surface of PECCs is covered with cells containing phenolic compounds in vacuoles; after OsO_4 prefixation, the phenolic is represented by black inclusion in vacuoles; (**b**) the phenolic can represent as large spots of irregular shape, which can completely fill the vacuole (double arrow); (c) the phenolic is deposited as small dots on the tonoplast of the vacuole PhC – phenolic-containing surficial cells

impregnate the specimen with resin; each resin is changed. If the air bubbles form, leave the container open until the air disappears.

6. The PECC embryogenic callus culture phenolic can be visualized staining with Toluidine Blue dye. For this purpose, PECC should be fixed in Karnovsky's fixative. On semi-thin

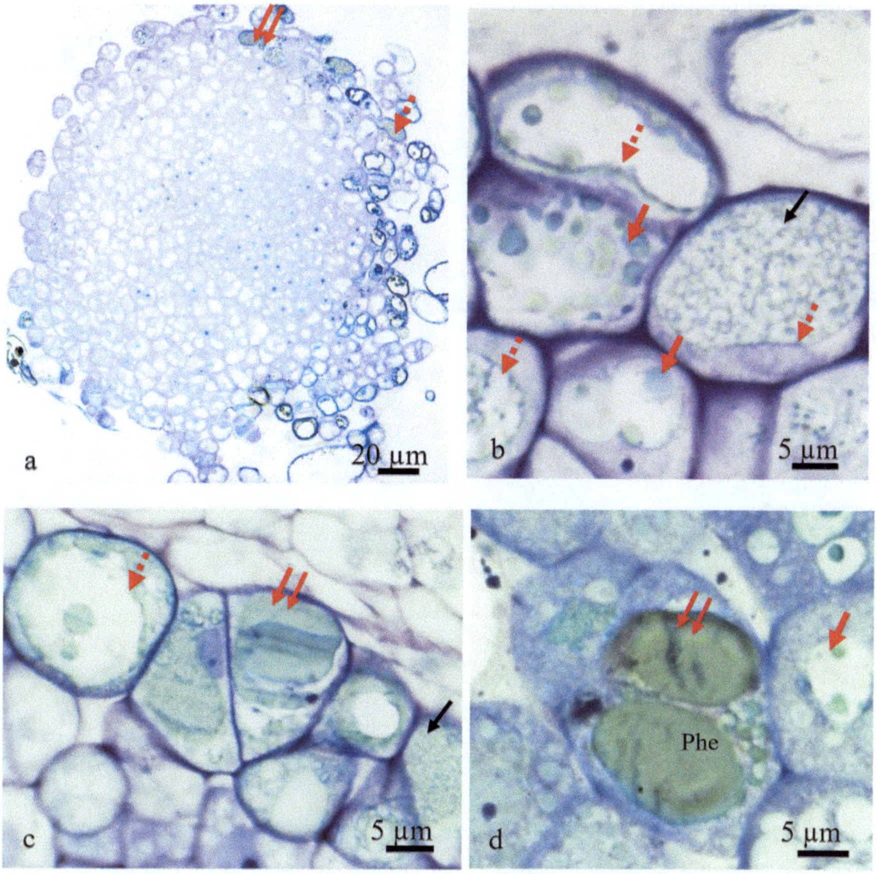

Fig. 2 Phenolic identification in proembryogenic cell complexes (PECCs) of morphogenic cell culture of buckwheat *Fagopyrum tataricum* (L.) Gaertn prefixed in Karnovsky's fixative and stained with Toluidine Blue O day; (**a**) on the histological section, the phenolic compound is represented as a bluish-greenish inclusion in vacuoles in the phenolic-containing superficial cells of PECC; (**b**) the phenolic can be deposited along the tonoplast like a thin line (broken arrow), as round inclusions (red arrow) in vacuoles and as small granular inclusions in the vacuole (black arrow); (c) (**d**) the phenolic can completely fill the vacuole (double arrow)

histological sections, the phenolic compounds are stained a greenish-blue color (Fig. 2).

7. The phenolic can be identified by prefixing the PECC with $FeCl_3$ and fixing in Karnovsky's fixative. The phenolic compounds are visualized as brownish inclusions in the vacuoles on the unstained semi-thin sections of epoxy resin fixed sections (Fig. 3).

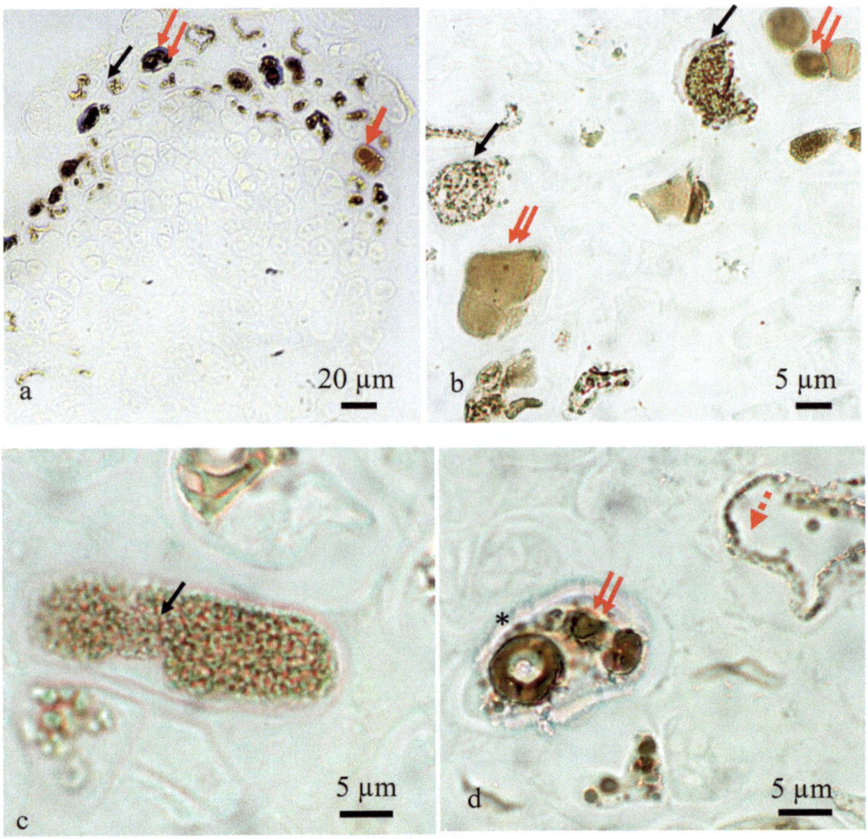

Fig. 3 Phenolic identification in proembryogenic cell complexes (PECCs) of morphogenic cell culture of buckwheat *Fagopyrum tataricum* (L.) Gaertn prefixed in Karnovsky's fixative and $FeCl_3$; (**a**) on histological section, the phenolic is represented as a brown inclusion in vacuoles in the phenolic-containing superficial cells of PECCs; (**b**) phenolic can completely fill the vacuole (double arrow) and can be represented as small granular inclusions in the vacuole (black arrow); (**c**) phenolic is represented as granular inclusions in the vacuole (black arrow); (**d**) a wide ring of phenolic along the tonoplast (star symbol)

Acknowledgments

The study was funded by the Ministry of Science and Higher Education of the Russian Federation (Minobrnauka) as part of the state assignments completed by KIBB FRC "Kazan Scientific Center of Russian Academy of Sciences" (No. 122011800137-0).

References

1. Horbowicz M, Wiczkowski W, Koczkodaj D, Saniewski M (2011) Effects of methyl jasmonate on accumulation of flavonoids in seedlings of common buckwheat (*Fagopyrum esculentum Moench*). Acta Biol Hung 62:265–278
2. Wahlsten A (2019) Phenolic compounds in *Fagopyrum* sp. grains (buckwheat); profile, bioactivity and effect of processing. Agronomprogrammet-Livsmedel Molekylära vetenskaper 3:8–34
3. Kreft S, Knapp M, Kreft I (1999) Extraction of rutin from buckwheat seeds and determination by capillary electrophoresis. J Agric Food Chem 47:4649–4652. https://doi.org/10.1021/jf990186p
4. Chandran H, Meen M et al (2020) Plant tissue culture as a perpetual source for production of industrially important bioactive compounds. Biotechnol Rep (Amst) 26:e00450. https://doi.org/10.1016/j.btre.2020.e00450
5. Betekhtin A, Rojek M, Jaskowiak J, Milewska-Hendel A, Kwasniewska J, Kostyukova Y et al (2017) Nuclear genome stability in long-term cultivated callus lines of *Fagopyrum tataricum* (L.) Gaertn. PLoS One 12(3):e0173537. https://doi.org/10.1371/journal.pone.0173537
6. Nielson A, Griffith W (1978) Tissue fixation and staining with osmium tetroxide: the role of phenolic compounds. J Histochem Cytochem 26(22):38–140. https://doi.org/10.1177/26.2.75221
7. Gutmann M (1995) Improved staining procedures for photographic documentation of phenolic deposits in semithin sections of plant tissue. J Microsc 179(3):277–281
8. Elwers S, Zambrano A, Ch R, Reinhard Lieberei R (2010) Histological features of phenolic compounds in fine and bulk cocoa seed (*Theobroma cacao* L.). J Appl Bot Food Qual 83(83):182–188
9. Bal A, Savory D (1980) Characterization of polyphenolcontaining dense cells. Experientia 36:1292–1294
10. Boufellous M, Lrhorfi A et al (2017) Phytochemical screening of a medicinal plant: *Lavandula stoechas* (Lamiaceae). J Pharmacogn Phytochem 6(2):56–62
11. Sumner M (2015) Epoxy resins for light and transmission electron microscopy. In: ECT Y et al (eds) Plant microtechniques and protocols, pp 83–101. https://doi.org/10.1007/978-3-319-19944-3_5
12. Hayat MA (2000) Principles and techniques of electron microscopy: biological applications, 4th edn. Cambridge University Press, Cambridge
13. Davies H (1999) Geeneral preparation of material and staining of sections. In: Electron microscopy methods and protocols, pp 1–13. https://doi.org/10.1385/1592592015
14. Trump BF, Smuckler EA, Benditt EP (1961) A method for staining epoxy sections for light microscopy. J Ultrastruct Res 5:343–348. https://doi.org/10.1016/S0022-5320(61)80011-7
15. Karnovsky MJ (1965) A formaldehyde-glutaraldehyde fixative of high osmolarity for use in electron microscopy. J Cell Biol 27:137a
16. Bal A, Savory D (1980) Characterization of polyphenol containing dense cells. Experientia 36:1292–1129

Chapter 5

Protoplast Isolation, Culture, and Regeneration in Common and Tartary Buckwheat

Magdalena Zaranek, Reneé Pérez-Pérez, Joanna Malec, and Ewa Grzebelus

Abstract

Techniques based on the use of plant protoplasts are a convenient model for better understanding and observing developmental changes in the cells. The establishment of research tools based on protoplasts consists of many steps needed for optimization. Here, we describe the culture of morphogenic callus (MC)- and hypocotyl-derived protoplasts of common (*Fagopyrum esculentum* Moench) and Tartary (*F. tataricum* (L.) Gaertn.) buckwheat. Protoplasts embedding in agarose matrix and application of plant hormones, including phytosulfokine (PSK), enable the development of protoplast cultures and plant regeneration.

Key words Agarose, Buckwheat, Driselase, Hypocotyls, In vitro cultures, Morphogenic callus, Phytosulfokine, Plant cells

1 Introduction

Single plant cells like protoplasts can be used in many fields, like genome editing, genetic manipulation, somatic hybridization, or the study of cell wall composition [1]. First, determine the protoplast source, enzyme mixture composition, and conditions of isolation effect on the quality of released protoplasts [2]. Next, protoplast culture requires optimization of the culture system and medium composition [3]. These are necessary for persistent mitotic divisions of protoplast-derived cells, aggregate formation, and their differentiation into plants [3]. Protoplasts of buckwheat species have been isolated from different explant sources with different successes [4–9]. According to the literature, authors have met problems in plant regeneration via protoplast cultures [4, 5, 9]. We have developed an efficient technique allowing plant

Magdalena Zaranek and Reneé Pérez-Pérez contributed equally with all other contributors.

Alexander Betekhtin and Artur Pinski (eds.), *Buckwheat: Methods and Protocols*, Methods in Molecular Biology, vol. 2791, https://doi.org/10.1007/978-1-0716-3794-4_5,

regeneration from morphogenic callus and hypocotyls of common and Tartary buckwheat. The positive response of protoplasts' development after their immobilization in agarose beads was demonstrated. Additionally, the KM medium [10] supplemented, i.e., with auxin, cytokinin, and especially PSK is crucial for buckwheat protoplast cultures, enabling cell division and further development. We noted that the regeneration potential of MC-derived protoplasts is considerably higher than hypocotyl-derived protoplasts [11]. This chapter describes protocols for buckwheat protoplast isolation, culture, and their regeneration into plants.

2 Materials

2.1 Solutions and Equipment

1. Seeds of Tartary buckwheat.
2. Seeds of common buckwheat.
3. Morphogenic callus of Tartary buckwheat.
4. Morphogenic callus of common buckwheat.
5. Sterile tea infuser.
6. 70% (*v/v*) ethanol.
7. 250 mL beakers.
8. I-FT sterilization solution: 0.1% (*v/v*) solution of fungicide, one drop of Tween-20.
9. II-FT sterilization solution: 20% (*w/v*) solution of chloramine T (sodium N-chlorotoluene-4-sulphonamide), 800 mg/L cefotaxime disodium, one drop of Tween-20.
10. I-FE sterilization solution: 0.2% (*v/v*) solution of fungicide, one drop of Tween-20.
11. II-FE sterilization solution: 20% (*w/v*) solution of chloramine T, 3000 mg/L cefotaxime disodium, one drop of Tween-20.
12. Sterile distilled water.
13. Sterile paper filter.
14. Sterile vacuum filter.
15. Vacuum pump.
16. Sterile filters (0.22 μm).
17. Glass Petri dishes (Ø9 cm × 2.5 cm).
18. Polystyrene Petri dishes (Ø9 cm × 1.4 cm).
19. Polystyrene Petri dishes (Ø6 cm × 1.2 cm).
20. Scalpel and blades.
21. Forceps.
22. Pasteur pipets.

23. Automatic pipette with tips.
24. Sterile 100 μm nylon sieve placed in funnel.
25. Sterile plastic 12 mL centrifuge tubs.
26. Fuchs-Rosenthal hemocytometer.
27. Gyratory shaker.
28. Scale.
29. Inverted microscope.
30. Centrifuge with swimming rotor.
31. Incubator.
32. Microwave.
33. Growth chamber.
34. Moss-coconut fiber substrate.

2.2 Media and Solutions

Pre-plasmolysis solutions, enzyme solutions, and liquid basal medium (BM) for protoplast cultures are adjusted to pH 5.6. The remaining media and solutions are adjusted to pH 5.8. BM, enzyme solutions, and low melting point agarose are filter sterilized (0.22 μm). The remaining solid media and solutions are sterilized in autoclave (0.1 MPa, 121 °C) and stored in room temperature. BM and enzyme solutions are stored in fridge.

1. FT germination medium: Murashige and Skoog (MS) medium with vitamins [12], 30 g/L sucrose, 7 g/L plant agar.
2. FE germination medium: MS medium with vitamins [12], 200 mg/L cefotaxime disodium, 30 g/L sucrose, 7 g/L plant agar.
3. Pre-plasmolysis solution (PSII): 0.5 M mannitol.
4. Pre-plasmolysis solution (PSII/F): 0.6 M mannitol, 5 mM $CaCl_2$.
5. Enzyme solution (E1): 1% cellulase Onozuka R10, 0.6% macerozyme R10, 0.1% driselase, 20 mM MES, 5 mM $MgCl_2 \times 6H_2O$, 0.6 M mannitol.
6. Enzyme solution (E2): 1.5% cellulase Onozuka R10, 0.15% driselase, 0.1% (*w/v*) pectolyase Y-23, 20 mM MES, 5 mM $MgCl_2 \times 6H_2O$, 0.6 M mannitol.
7. Enzyme solution (E3): 1% (*w/v*) cellulase Onozuka R10, 0.6% (*w/v*) macerozyme R10, 0.25% (*w/v*) driselase, 20 mM 2-(N-Morpholino) ethanesulfonic acid (MES), 5 mM $MgCl_2 \times 6H_2O$, 0.6 M mannitol.
8. Enzyme solution (E4): 1% cellulase Onozuka R10, 0.1% pectolyase Y-23, 20 mM MES, 5 mM $MgCl_2 \times 6H_2O$, 0.6 M mannitol.
9. Suc-2/MES: 0.6 M sucrose, 1 mM MES.

10. Suc/MES: 0.5 M sucrose, 1 mM MES.
11. Washing solution (W5): 154 mM NaCl, 125 mM $CaCl_2 \times 2H_2O$, 5 mM KCl, 5 mM glucose.
12. Liquid basal medium for protoplast cultures (BM): KM macro- and microelements and organic acids [10], 100 mg/L myo-inositol, 10 mg/L thiamine, 1 mg/L pyridoxine, 1 mg/L nicotinic acid, 0.25 g/L N-Z-amine A, 1 mg/L 6-benzylaminopurine (BAP), 2 mg/L 1-naphthalenacetic acid (NAA), 0.4 M glucose.
13. Low melting point agarose (LMPA): 1.2% SeaPlaque agarose in BM. (**Note 1:** Heat in the microwave to dissolve agarose, and filter after cooling.)
14. 1×10^{-4} M working solution of phytosulfokine (PSK).
15. 8 mg/L working solution of putrescine (PUT).
16. 0.25 mg/L working solution of chloropyridin phenylurea (CPPU).
17. 50% (*w/v*) working solution of polyvinylpyrrolidone (PVP).
18. 200 mg/L working solution of cefotaxime disodium.
19. Callus multiplication medium (CM): MS medium with vitamins [12], 2 g/L N-Z-amine A, 2 mg/L 2,4-dichlorophenoxyacetic acid (2,4-D), 0.2 mg/L kinetin (KIN), 100 nM PSK, 30 g/L sucrose, 3 g/L phytagel.
20. Regeneration medium (RM_MS3): MS medium [12], 3 mg/L BAP, 1 mg/L thidiazuron (TDZ), 30 g/L sucrose, 3 g/L phytagel.
21. Regeneration medium (RM_MS4): MS medium [12], 2 mg/L BAP, 1 mg/L KIN, 0.0025% PVP, 30 g/L sucrose, 3 g/L phytagel.
22. Rooting medium: MS medium with vitamins [12], 30 g/L sucrose, and 3 g/L phytagel.

3 Methods

3.1 Hypocotyl-Derived Protoplasts of Common Buckwheat

3.1.1 Seed Sterilization and Germination

1. For hypocotyl development, keep the seeds of common buckwheat overnight in distilled water. Next day, remove the seed coat. (**Note 2**: Remove the seed coat by cutting of the seam using magnifier and scalpel. Try to do not damage endosperm and embryo.)
2. For surface sterilization, place the seeds in tea infuser and soak in ethanol for 30 s, then transfer to a beaker with 100 ml of I-FE sterilization solution, and put on a gyratory shaker (160 rpm) for 30 min at RT. Dip the tea infuser in ethanol for 30 s, next transfer to 100 ml of II-FE sterilization solution,

and put on a gyratory shaker for 30 min. Under laminar flow, wash the seeds three times with sterile distilled water for 5 min each, and soak overnight in sterile distilled water at (room temperature) RT (use aluminum foil to wrap the beaker).

3. The next day, repeat sterilization in II-FE solution, and wash the seeds three times in sterile distilled water for 5 min each. Under laminar flow, transfer seeds from tea infuser on a sterile filter paper and air-dry for 15 min.
4. To grow hypocotyls for protoplast isolation, transfer six seeds per polystyrene Petri dish (Ø9 cm) with FE germination medium. Seal the Petri dishes with parafilm, and incubate in horizontal position in the dark at 26 ± 1 °C for 10 days. (**Note 3**: Do not place more than six seeds per Petri dish to avoid contaminations.)

3.1.2 Protoplast Isolation and Culture

1. Transfer 1 g of 10-day-old hypocotyls to a glass Petri dish with 10 mL of PSII solution, then cut into 1-cm-long pieces, and then each piece cut longitudinally. Incubate for 1 h in the dark at RT. (**Note 4:** Take only hypocotyls without root and cotyledons.)
2. Remove PSII solution by Pasteur pipette, and add 10 mL of E1 enzyme solution. Then incubate the material overnight (for 16 h) with gentle shaking (50 rpm) at RT in the dark. (**Note 5**: To prevent endogenous bacterial contaminations to the Petri dish with E1 solution, 800 mg/L cefotaxime disodium can be added.)
3. Check the quality of released protoplasts under inverted microscope. (**Note 6:** If you note high amount of undigested tissue, higher concentration of driselase (0.15–0.25%) can be applied.)
4. Pour the protoplast suspension to centrifuge tubes through nylon sieves, and centrifuge at 1000 rpm for 5 min.
5. Remove the supernatant and resuspend pellet up to 8 ml of Suc/MES, then very slowly add 2 mL of W5 solution to form a gradient, and then centrifuge at 1200 rpm for 10 min.
6. Very gently collect protoplast ring with the Pasteur pipette to new centrifuge tube, add up to 10 mL W5 solution, and then centrifuge at 1000 rpm for 5 min.
7. Remove the supernatant and again resuspend pellet up to 10 mL in BM medium, and centrifuge at 1000 rpm for 5 min.
8. Resuspend the pellet in 1 mL of BM, and determine the yield of protoplasts using a Fuchs-Rosenthal hemocytometer. Adjust working density to 5×10^5 cells/ml by adding BM medium if necessary.
9. Embed the protoplasts in LMPA by combining equal volume of protoplast suspension at the density of 5×10^5 cells/ml with

LMPA. Then drop the suspension at a rate of four 50 μL beads per Petri dish (Ø6 cm). After 15 min of agarose solidification, add BM. Supplement the BM medium with 100 nM PSK, 8 mg/L PUT, and 0.25 mg/L CPPU. (**Note** 7: To prevent bacterial contamination, 800 mg/L cefotaxime disodium can be add to the BM.) Seal the Petri dishes with parafilm.

10. Incubate protoplast cultures for 60 days at 26 ± 1 °C in the dark. After 10 days of culture, remove BM medium and add fresh one with the same additional supplements.

3.1.3 Protoplast-Derived Callus Multiplication and Plant Regeneration

1. Approximately after 2 months of protoplast cultures, transfer the overgrown agarose drops by microcallus to CM medium, and incubate at 26 ± 1 °C in the dark. Subculture them every 3 weeks on the same CM medium.
2. Pick the calli that double its mass, and cultivate them on glass Petri dishes with regeneration medium RM_MS4. Maintain in a growth room at 28 ± 2 °C with a 16/8 h (light/dark) photoperiod, under light intensity of 55 μmol/m^2/s. Subculture callus every 3 weeks no fresh RM_MS4 medium. Seal the Petri dishes with nonwoven tape.
3. Transfer shoots to sterile ventilated vessels containing rooting medium, and maintain them in a climate room at 25 ± 1 °C with a 16/8 h (light/dark) photoperiod (light intensity 55 μmol/m^2/s).
4. Transfer fully developed plants into a moss-coconut fiber substrate, and maintain in greenhouse at 25 ± 1 °C with 16/8 h (light/dark) photoperiod and light intensity 90 μmol/m^2/s.

3.2 MC-Derived Protoplasts of Common Buckwheat

3.2.1 Protoplast Isolation and Culture

1. Transfer 2 g of 12-day-old MC (Fig. 1a) to a glass Petri dish with 5 mL of PSII/F pre-plasmolysis solution, and ground the tissue with a sterile scalpel. Then, add 5 mL more of PSII/F, and incubate for 1 h at RT in the dark.
2. Carefully remove the PSII/F solution with a Pasteur pipette without taking the tissue, and add 10 ml of E2 enzyme solution. Seal the Petri dishes with parafilm, and shake them for 14–16 h in the dark (30–50 rpm, RT).
3. After enzyme treatment, the tissue should be completely digested as shown in Fig. 1c. Check the quality of the released protoplasts under the inverted microscope. They should have round shape (Fig. 1e).
4. Filter the suspension into a sterile tube using a nylon sieve, and centrifuge for 5 min (1000 rpm, RT). Remove the supernatant without disturbing the pellet, and resuspend it in 8 mL of Suc/MES solution. Slowly, drop 2 mL of W5 solution down the tube walls to form a gradient (Fig. 1g). Centrifuge for 10 min (1200 rpm, RT).

Fig. 1 Steps for protoplasts isolation and culture from MC of (**a**) *F. esculentum* and (**b**) *F. tataricum.* (**c, d**) Digested callus after enzyme treatment; (**e, f**) protoplasts released after 16 h of enzymatic maceration observed under microscope; (**g**) gradient between protoplasts resuspended in Suc/MES (down) and W5 (up) solution; (**h**) ring of viable protoplasts at the interphase between Suc/MES and W5 (blackhead arrows); (**i**) embedding of protoplasts in LMPA (1-day-old cultures); (**j**) microcallus overgrowing in agarose beads (40-day-old cultures); (**k**) multiplied morphogenic callus on CM medium (2-month-old callus cultures); development on CM medium; (**l**) embryogenic cell masses (yellowhead arrows) of *F. esculentum*; (**m**) embryogenic complexes (yellow head arrows) and (**n**) somatic embryos of *F. tataricum. Fe: F. esculentum; Ft: F. tataricum; LMPA: low melting point agarose.* Scale bars: 1 mm (**a, b, l, m, n**), 100 μm (**e, f**)

5. At the gradient interface, a ring of viable protoplasts should be positioned (Fig. 1h). Collect the protoplast with a Pasteur pipette and transfer them to a new sterile tube. Add up to 10 mL W5 solution and centrifuge at 1000 rpm for 5 min.
6. Remove the supernatant and resuspend the pellet in 10 mL of BM medium. Centrifuge for 5 min at 1000 rpm.

7. Remove completely the supernatant with a Pasteur pipette without disturbing the pellet, and resuspend it in 1 mL of BM medium.
8. Determine the yield of protoplasts using a Fuchs-Rosenthal hemocytometer, and adjust working density to 8×10^5 cells/ ml by adding BM medium if necessary.
9. Embed the protoplasts in LMPA combining equal volume of protoplast suspension at the density of 8×10^5 cells/ml with LMPA. Make four drops (about 50 μL) in polystyrene Petri dishes (Ø6 cm), and let them solidify (Fig. 1i). Then add 4 ml of BM medium supplemented with 100 nM of PSK, 8 mg/L PUT, and 0.25 mg/L CPPU. Seal with parafilm and incubate in the dark at 26 ± 2 °C. After 10 days of culture, remove BM medium and add fresh one with the same additional supplements.

3.2.2 Protoplast-Derived Callus Multiplication and Plant Regeneration

1. When the beads are overgrown with callus, as shown in Fig. 1j, transfer them to Petri dishes (Ø6 cm) with CM medium to multiply the callus (Fig. 1k). Incubate in the dark at 26 °C ± 2 ° C, and subculture them every 3 weeks on the same medium. This step could take around 2 months.
2. When embryogenic masses (Fig. 1l) or somatic embryos appear in the callus, transfer them to RM_MS3 regeneration medium. Seal the Petri dishes with nonwoven tape, and incubate in climate room at 28 ± 2 °C with a 16/8 h (light/dark) photoperiod and light intensity 55 μmol/m^2/s. Transfer to fresh medium every 3 weeks.
3. Transfer the emerging shoots to sterile ventilated vessels containing rooting medium, and maintain them in climate room at 25 ± 1 °C with a 16/8 h (light/dark) photoperiod and light intensity 55 μmol/m^2/s.
4. Once full plants have developed, transfer them into moss-coconut fiber substrate, and maintain in greenhouse at 25 ± 1 °C with 16/8 h (light/dark) photoperiod and light intensity 90 μmol/m^2/s.

3.3 Hypocotyl-Derived Protoplasts of Tartary Buckwheat

3.3.1 Seed Sterilization and Germination

1. Place Tartary buckwheat seeds in tea infuser, and then dip in ethanol for 30 s. Sterilize seeds in a beaker using 100 mL of I-FT sterilization solution, and place on a gyratory shaker (160 rpm) for 30 min at RT, and then dip the seeds in ethanol for 30 s.
2. Next transfer the tea infuser with seeds to the second beaker with 100 ml of II-FT sterilization solution, and shake (160 rpm) for 30 min, and then dip the seeds in ethanol for 30 s.

3. Under the laminar flow, wash the seeds three times in sterile distilled water for 5 min each, and leave in beaker with sterile distilled water overnight at RT (wrapped in aluminum foil).
4. Under sterile conditions, repeat washing the seeds with sterile distilled water as described above, and air-dry the seeds on a sterile filter paper for 15 min.
5. To germinate hypocotyls, transfer about eight seeds into polystyrene Petri dish (Ø9 cm) containing FT germination medium, seal the Petri dishes with parafilm, and culture them in horizontal position at 26 ± 1 °C in the dark for 10 days. (**Note 8**: For 1 g hypocotyls, two Petri dishes with seeds are enough.)

3.3.2 Protoplast Isolation and Culture

1. Place 1 g of 10-day-old hypocotyls in a glass Petri dish with 10 mL of PSII pre-plasmolysis solution, cut into pieces of 1 cm in length, and then cut them longitudinally. Place the Petri dish in the dark for 1 h at RT.
2. Remove PSII solution with Pasteur pipette, and add 10 mL of E3 enzyme solution. Incubate material on gyratory shaker (40–50 rpm) overnight (16 h) in the dark at RT. (**Note 9**: To avoid endogenous bacterial contamination, 800 mg/L cefotaxime disodium can be added.)
3. Check quality of released protoplasts under inverted microscope. (**Note 10:** Released protoplasts should be round in shape. If the tissue is not digested properly and the amount of protoplasts is low, put the dish again on gyratory shaker for 1 h more.)
4. Filter the digested material into a sterile centrifuge tube using a funnel and nylon sieve. Centrifuge the suspension at 1000 rpm for 5 min.
5. Remove supernatant and resuspend pellet up to 8 ml of Suc-2/MES solution, then slowly overlay 2 mL of W5 solution, and centrifuge at 1200 rpm for 10 min.
6. With a Pasteur pipette, very carefully collect protoplasts localized in the interphase and transfer to a new centrifuge tube, add up to 10 mL of W5 solution, and centrifuge at 1000 rpm for 5 min.
7. Remove supernatant, resuspend the pellet in BM medium, and centrifuge as described above.
8. Remove supernatant and resuspend pellet up to 1 mL.
9. Determine the yield of isolated protoplasts using a Fuchs-Rosenthal hemocytometer, and adjust the working density to 5×10^5 cells/ml by adding BM medium if necessary.

10. Once the protoplast concentration is adjusted, add the same volume of LMPA, then place four 50 μL aliquot drops of the LMPA-protoplasts mixture onto the bottom of a Petri dish (Ø6 cm), and incubate (for approx. 15 min) until the agarose solidifies.
11. Then add 4 mL of BM and supplement with 100 nM PSK, 8 mg/L PUT, and 0.25 mg/L CPPU. Seal the Petri dishes with parafilm, and incubate at 26 °C ± 1 °C in the dark for 2 months. (**Note 11:** During intensive development of protoplasts, about 10 days around the agarose bead coating of metabolites can be observed. Therefore, to prevent browning of the culture due to accumulation of phenolic, 0.025–0.05% PVP can be added. Additionally, the medium can be replaced after 10 and 25 days of culture.)
12. Remove BM medium and add fresh one with the same supplements after 10 days of culture.

3.3.3 Protoplast-Derived Callus Multiplication and Plant Regeneration

1. After 2 months of culture, remove the liquid BM medium using a Pasteur pipette, and transfer the microcallus on Petri dish with solid CM medium. Incubate in the dark at 26 ± 1 °C for 3 weeks. After this time, transfer callus clumps to new CM medium.
2. To plant regeneration, transfer callus clumps onto the glass Petri dishes with regeneration medium RM_MS4, seal the Petri dishes with nonwoven tape, and culture in a growth room at 28 ± 2 °C with a 16/8 h (light/dark) photoperiod, under light intensity of 55 μmol/m^2/s subculturing every 3 weeks. (**Note 12**: In our study, the regeneration of hypocotyl-derived protoplasts of Tartary buckwheat was not noted. We suppose that it is due to the applied genotype or polysomatic nature of these plants [13].)

3.4 Protoplasts Derived from MC of Tartary Buckwheat

3.4.1 Isolation and Culture of Protoplasts

1. Transfer 1 g of 10-day-old calli of Tartary buckwheat (Fig. 1b) to a glass Petri dish with 10 mL of PSII/F pre-plasmolysis solution, and incubate for 1 h at RT in the dark.
2. Carefully remove the PSII/F solution with a Pasteur pipette without taking the tissue, and add 10 mL of E4 solution. Seal the Petri dishes with parafilm, and shake them for 14–16 h in the dark (30–50 rpm, RT). (**Note 13:** To remove PSII/F solution, a centrifugation step (5 min, 1000 rpm, RT) can be done to separate solution and tissue (at the bottom of the tube).)
3. After maceration, the tissue should be completely digested as shown in Fig. 1d. Check the quality of the released protoplast under the microscope. They should have round shape (Fig. 1f).

4. Filter the suspension into a sterile tube using a nylon sieve, and centrifuge for 5 min (1000 rpm, RT). Remove the supernatant without disturbing the pellet, and resuspend it in 8 mL of Suc/MES solution. Slowly, drop 2 mL of W5 solution down the tube walls to form a gradient (Fig. 1g). Centrifuge for 10 min (1200 rpm, RT).
5. At the gradient interface, a ring of viable protoplasts should be positioned (Fig. 1h). Collect the protoplast with a Pasteur pipette, and transfer them to a new sterile tube. Add up to 10 mL of W5 solution, and centrifuge at 1000 rpm for 5 min.
6. Remove the supernatant and resuspend the pellet in 10 mL of BM medium. Centrifuge for 5 min at 1000 rpm.
7. Remove completely the supernatant with a Pasteur pipette without disturbing the pellet, and resuspend it in 1 mL of BM medium.
8. Determine the yield of protoplasts using a Fuchs-Rosenthal hemocytometer, and adjust working density to 8×10^{-5} cells/ml by adding BM medium if necessary.
9. Embed the protoplasts in LMPA combining equal volume of protoplasts suspension at the density of 8×10^{-5} cells/ml with LMPA. Make four drops (about 50 μL) in polystyrene Petri dishes (Ø6 cm), and let them solidify (Fig. 1i). Then add 4 mL of BM medium supplemented with 100 nM of PSK, 8 mg/L PUT, and 0.25 mg/L CPPU. Seal with parafilm and incubate in the dark at 26 °C ± 2 °C. After 10 days of culture, remove BM medium and add fresh one with the same additional supplements. (**Note 14:** During intensive development of protoplasts about 10–20 days, coating of metabolites can be observed around the agarose beads. Therefore, to prevent browning of the culture due to accumulation of phenolic, 0.025% of PVP can be added. Additionally, the medium can be replaced after 10 and 25 days of culture.)

3.4.2 Callus Multiplication and Plant Regeneration of Tartary Buckwheat

1. When the beads are overgrown with calli, as shown in Fig. 1j, transfer them to Petri dishes (Ø6 cm) with CM medium to multiply the calli (Fig. 1k). Incubate in the dark at 26 ± 2 °C, and subculture them every 3 weeks on the same medium. This step could take around 2 months.
2. When pro-embryogenic cell complexes (Fig. 1m) or somatic embryos (Fig. 1n) appear on the callus, transfer them to RM_MS4 regeneration medium. Seal the Petri dishes with nonwoven tape, and incubate in climate room at 28 ± 2 °C with a 16/8 h (light/dark) photoperiod and light intensity of 55 $\mu mol/m^2/s$. Transfer to fresh medium every 3 weeks.
3. Transfer the emerging shoots to sterile ventilated vessels containing rooting medium, and maintain them in climate room at

25 ± 1 °C with a 16/8 h (light/dark) photoperiod and light intensity of 55 μmol/m^2/s.

4. Once full plants have developed, transfer them into moss-coconut fiber substrate, and maintain in greenhouse at 25 ± 1 °C, 16/8 h (light/dark) photoperiod, and light intensity of 90 μmol/m^2/s.

Acknowledgments

This work was supported by the National Science Centre, Poland: Research projects SONATA BIS 10 (2020/38/E/NZ9/00033 awarded to AB) and OPUS (2020/37/B/NZ9/01499 awarded to AB).

References

1. Grosser JW, Ćalović M, Louzada ES (2010) Protoplast fusion technology – somatic hybridization and cybridization. In: Plant Cell Culture, pp 175–198. https://doi.org/10.1002/9780470686522.ch10
2. Reed KM, Bargmann BOR (2021) Protoplast regeneration and its use in new plant breeding technologies. Front Genome Ed 3:734951. https://doi.org/10.3389/fgeed.2021.734951
3. Davey MR, Anthony P, Power JB, Lowe KC (2005) Plant protoplasts: status and biotechnological perspectives. Biotechnol Adv 23(2): 131–171. https://doi.org/10.1016/j.biotechadv.2004.09.008
4. Adachi T, Yamaguchi A, Miike Y, Hoffmann F (1989) Plant regeneration from protoplasts of common buckwheat (*Fagopyrum esculentum*). Plant Cell Rep 8(4):247–250. https://doi.org/10.1007/BF00778544
5. Lachmann S, Adachi T (1990) Callus regeneration from hypocotyl protoplast of tartary buckwheat (*Fagopyrum tataricum* Gaertn.). Fagopyrum 10:62–64
6. Holländer-Czytko H, Amrhein N (1983) Subcellular compartment of shikimic acid and phenylalanine in buckwheat cell suspension cultures grown in the presence of shikimate pathway inhibitors. Plant Sci Lett 29(1): 89–96. https://doi.org/10.1016/0304-4211(83)90027-5
7. Sakamoto S, Matsui K, Oshima Y, Mitsuda N (2020) Efficient transient gene expression system using buckwheat hypocotyl protoplasts for large-scale experiments. Breed Sci 70(1): 128–134. https://doi.org/10.1270/jsbbs.19082
8. Woo S-H, Adachi T, Jong SK, Campbell CG (1999) Isolation of protoplasts from viable sperm cells of common buckwheat (Fagopyrum esculentum Moench.). Can J Plant Sci 80:583–585
9. Gumerova EA (2004) Realisation of the morphogenic potential of common buckwheat (Fagopyrum esculentum Moench.) hypocotyls depending on the method of regeneration KIBB. KSC Russian Academy of Science, Kazan
10. Kao KN, Michayluk MR (1975) Nutritional requirements for growth of *Vicia Hajastana* cells and protoplasts at a very low population density in liquid media. Planta 126(2): 105–110. https://doi.org/10.1007/Bf00380613
11. Zaranek M, Pérez-Pérez R, Milewska-Hendel-A, Grzebelus E, Betekhtin A (2023) Efficient and rapid system of plant regeneration via protoplast cultures of *Fagopyrum esculentum* Moench. PCTOC 154:673. https://doi.org/10.1007/s11240-023-02542-2
12. Murashige T, Skoog F (1962) A revised medium for rapid growth and bio assays with tobacco tissue cultures. Physiol Plantarum 15(3):473–497. https://doi.org/10.1111/j.1399-3054.1962.tb08052.x
13. Betekhtin A, Rojek M, Jaskowiak J, Milewska-Hendel A, Kwasniewska J, Kostyukova Y, Kurczynska E, Rumyantseva N, Hasterok R (2017) Nuclear genome stability in long-term cultivated callus lines of *Fagopyrum tataricum* (L.) Gaertn. PLoS One 12:3. https://doi.org/10.1371/journal.pone.0173537

Chapter 6

Immunohistochemical Detection of the Wall Components on the Example of Shoot Apical Meristem of *Fagopyrum esculentum* and *Fagopyrum tataricum*

Anna Milewska-Hendel, Ewa Kurczyńska, and Kamila Godel-Jędrychowska

Abstract

Immunohistochemistry is a method that allows the detection of individual components of cell walls in an extremely precise way at the level of a single cell and wall domains. The cell wall antibodies detect specific epitopes of pectins, arabinogalactan proteins (AGP), hemicelluloses, and extensins. The presented method visualization of the selected pectic and AGP epitopes using antibodies directed to wall components is described. The method of the analysis of the chemical composition of the wall is present on the example of the shoot apical meristems of *Fagopurum esculentum* and *Fagopyrum tataricum*. Recommended protocols for immunostaining and examination on fluorescence microscopy level are presented.

Key words Arabinogalactan proteins, Cell wall, Fagopyrum, Pectins

1 Introduction

The cell wall is an integral part of each plant cell. Due to the functions that the cell wall performs, i.e., active participation in cell metabolism, reaction to internal and external stimuli, as well as biotic and abiotic factors, the chemical composition of the cell wall is an excellent marker of the state of the plant cell during cell differentiation. Moreover, the wall is a source of signals regulating plant development [1–4].

The structure and chemical composition of the cell wall are complex. The main components that build the wall are cellulose as a structural material, pectins, hemicelluloses, and proteins, including hydroxyproline-rich glycoproteins (HRGP; extensin), the arabinogalactan proteins (AGP), the glycine-rich proteins (GRPs), and the proline-rich proteins (PRPs) [2].

Alexander Betekhtin and Artur Pinski (eds.), *Buckwheat: Methods and Protocols*, Methods in Molecular Biology, vol. 2791, https://doi.org/10.1007/978-1-0716-3794-4_6,

The chemical composition of cell walls can be analyzed using immunohistochemistry. It is a method allowing the identification of specific antigens within walls and cells with antigen-specific antibodies, which bind a protein in a section and use fluorochromes to visualize the presence of antibodies, which allows analysis of the localization of individual epitopes [2].

Due to the specific structure, i.e., the presence of polysaccharides, the cell wall requires a more complicated procedure to obtain antibodies directed to individual epitopes, and the preparation of antibodies directed to such wall components is not direct [2]. Briefly, this involves the coupling of a defined oligosaccharide to an immunogenic protein such as BSA (bovine serum albumin), and the generated probes are mostly rat hybridoma or phage display monoclonal antibodies that allow the generation of a series of monoclonal antibodies that recognize diverse carbohydrate structures within plant cell walls. We recommend the Paul Knox Cell Wall Lab at the University of Leeds for more information. Some examples of antibodies used to analyze the wall chemical components are presented below (Table 1).

2 Materials

2.1 Culture

1. Plastic pots.
2. Garden soil.
3. Vermiculite.
4. Sterile culture dishes.

2.2 Fixation

1. 0.1 M PBS (phosphate buffered saline): dissolve 8 g NaCl, 0.2 g KCl, 0.24 g KH_2PO_4, and 1.44 g Na_2HPO_4 in 1 L of distilled water, and adjust the pH to 7.2 (*see* **Note 1**).
2. 3% glutaraldehyde (GA) and 1% paraformaldehyde fixative: use 6 mL of 25% GA and 0.5 g PFA to prepare 50 mL fixative solution in 0.1 M PBS (v/v; *see* **Note 2**).
3. Glass beaker.
4. Filter paper.
5. 2 mL Eppendorf tubes.
6. Stereomicroscope Nikon microscope (Japan).
7. Forceps.
8. Razor blades.
9. pH meter.
10. Desiccator.
11. Aluminum foil.
12. Heating plate.

Table 1
List of primary rat monoclonal antibodies

Antibody	Epitope	References
Pectic polysaccharides		
Homogalacturonan (HG) – Methyl-esterification degree		
LM19 JIM5	Non-methyl-esterified, partially methyl-esterified HG	[5] [6]
LM20 JIM7	Methyl-esterified HG	[5] [6]
Rhamnogalacturonan-I – Side chains		
LM5	Linear tetrasaccharide in (1–4)-β-D-galactans	[7]
LM6	Linear pentasaccharide in (1–5)-α-L-arabinans	[8]
LM13	Longer stretches of 1,5-linked arabinosyl residues	[9]
Arabinogalactan-protein (AGP) glycan		
JIM8	Arabinogalactan	[10]
JIM13	Arabinogalactan/arabinogalactan protein	[11]
JIM16	Arabinogalactan/arabinogalactan protein	[11]
MAC207	Arabinogalactan protein	[12]

2.3 Dehydration

1. Ethanol series solutions (10%, 30%, 50%, 70%, 90%, in dH_20 (v/v) and 100% ethanol; *see* **Note 3**).

2.4 Infiltration and Embedding

1. LR White resin.
2. Solution 3:1 (v/v) 100% ethanol/LR White resin: 15 mL of ethanol mixed with 5 mL LR White resin.
3. Solution 1:1 (v/v) 100% ethanol/LR White resin: 10 mL of ethanol mixed with 10 mL LR White resin.
4. Solution 1:3 (v/v) 100% ethanol/LR White resin: 5 mL of ethanol mixed with 15 mL LR White resin.
5. Gelatin capsules (*see* **Note 4**).
6. Preparation needles.
7. Forceps.

2.5 Tissue Sectioning

1. Ultramicrotome.
2. Histo 45° diamond knife Diatome.
3. Poly-L-lysine microscope slides.
4. Razor blades.
5. Distilled water (dH_2O).

6. Toluidine blue O solution (TBO; 0.05% solution in dH2O; dissolve 0.05 g TBO in 100 mL dH2O).
7. Slide storage box.

2.6 Immunostaining

1. Blocking buffer: 2% BSA (bovine serum albumin), 2% FCS (fetal calf serum) in 0.1 M PBS buffer. Dissolve 1 g BSA and 1 mL FCS in 50 mL PBS.
2. Primary antibodies working solutions: dilution of 1:20 in blocking buffer (6 μL antibody per 114 μL blocking buffer per one outline on slide).
3. Secondary antibody goat anti-rat IgG H + L, fluorochrome AlexaFluor®488 conjugated: dilution of 1:100 in blocking buffer (1.2 μL antibody per 118.8 μL blocking buffer per one outline on a slide; *see* **Note 5**).
4. Fluorescence Brightener 28 (FB28, calcofluor); 0.01% solution of FB28 in PBS: dissolve 0.01 g of FB28 in 100 mL of PBS, shake until dissolved, and then filter through a 0.22 μm syringe filter (*see* **Note 6**).
5. PAP pen.
6. Humid slide chamber (*see* **Note** 7).
7. Ultrapure water.
8. Fluoromount.
9. Coverslip.
10. Epifluorescence microscope equipped with UV and blue light illumination filters and barrier filters for green emission.

3 Methods

3.1 Culture

1. Sow *Fagopyrum esculentum* and *Fagopyrum tataricum* seeds in two separate moist plastic pots. Put ten seeds into one moist plastic pot with garden soil and vermiculite in a ratio of 1:1.
2. Leave seeds to germinate in a greenhouse at 25 ± 1 °C and a 16 h photoperiod with a light intensity of 40 μmol/s/m^2.
3. Collect the plants between 7 and 11 days (Fig. 1a).
4. Cut the apical part by the razor blade (not more than 5 mm long) of plants *Fagopyrum esculentum* and *Fagopyrum tataricum* using the stereomicroscope, and immediately put them into fixative solution using fine forceps (Fig. 1b).

3.2 Fixation

1. Carefully transfer the sample into a 2 mL Eppendorf tube with 1.5 mL of the 3% PFA and 1% GA fixative (*see* **Note 8**).

   Fixative preparation

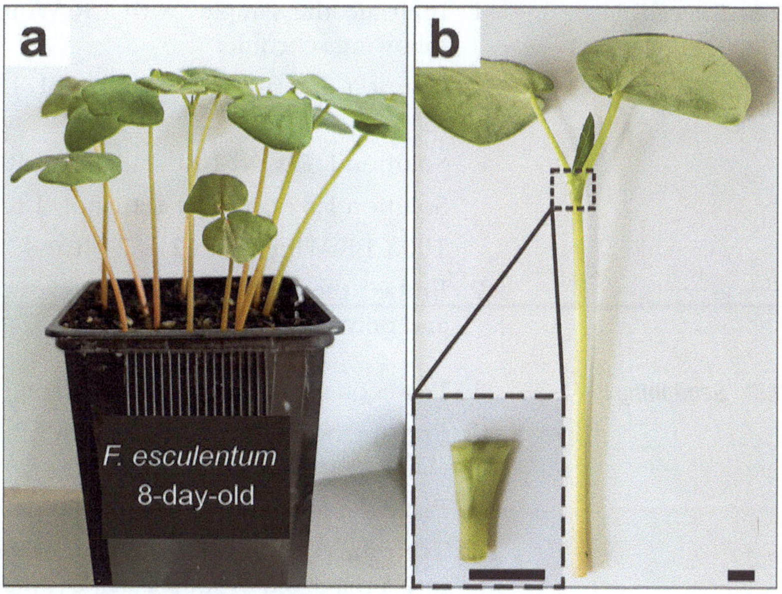

Fig. 1 Preparation of material for the procedure. (**a**) Seedlings of *Fagopyrum esculentum* at the appropriate stage to collect meristems for immunohistological analysis. (**b**) Analyzed seedling: the square indicates the area to be sampled for analysis. (**b inset**) Fragment containing shoot apical meristem, which should be fixed. Scale bars – 5 mm

   A. Prepare a 75–100 mL glass beaker.
   B. Pour approx. 40 mL of 0.1 M PBS in a glass beaker.
   C. Add 0.5 g of PFA into a 40 mL 0.1 M PBS beaker and cover with aluminum foil.
   D. Place on the heating plate (temperature 50–55 °C) and dissolve for approx. 1–1.5 h (*see* **Note 9**). Work with attention under a well-functioning fume hood. Use gloves.
   E. Cool the solution and filter through a filter paper.
   F. Add 6 mL of GA.
   G. Add PBS to the final volume (50 mL).
   H. Adjust to pH 7.2 (*see* **Note 10**).
2. Deaerate the samples inside the desiccator chamber under the vacuum (*see* **Note 11**).
3. Leave the sample with 3% PFA and 1% GA fixative overnight at 4 °C.

3.3 Dehydration

1. Rinse the samples with 0.1 M PBS for 3 × 10 min.
2. Dehydrate the sample in increasing ethanol from 10% to 100% series. Keep the samples in each concentration for 2 × 30 min at room temperature (*see* **Note 12**).

3.4 Embedding

1. Infiltrate the samples with LR White resin according to the following schedule:

   Solution 3:1 (v/v) 100% ethanol/LR White resin – 24 h at 4 °C

   Solution 1:1 (v/v) 100% ethanol/LR White resin – 24 h at 4 °C

   Solution 1:3 (v/v) 100% ethanol/LR White resin – 24 h at 4 °C

   100% LR White resin 2 × 24 h at 4 °C

2. Embed the samples in gelatin capsules with LR White resin, and polymerize them for 24 h at 57 °C (*see* **Note 13**).

3.5 Sectioning

1. Power on the ultramicrotome and prepare it to work (Fig. 2a–d).
2. Prepare the samples embedded in LR White resin for sectioning (Fig. 3a; *see* **Note 14**).
3. Place the block on the sample holder (Fig. 3b; *see* **Note 15**).
4. Trim the block around the area where the material is located into a trapezoid or square shape (block face) around the sample using a razor blade (Fig. 3c–d; *see* **Note 16**).
5. Place the specimen holder with the trimmed block (long edge of the trapezoid down) on the specimen arm of the ultramicrotome (Fig. 3e; *see* **Note 17**).
6. Place the knife in the knife holder (Fig. 3f; *see* **Note 18**).
7. Fill the knife boat with dH_2O until it does *not* touch the knife's edge, and move the knife to a distance of approximately 3 mm from the block (Fig. 3g).
8. Properly orient the block with the embedded material in relation to the knife so as to obtain longitudinal sections. This means that the material in the block should be parallel to the back of the knife.
9. Fill the knife boat with dH_2O until it touches the knife's edge, and move the knife as close to the face of the block as possible (Fig. 3h–j; *see* **Note 19**).
10. Cut the sample into 1–1.5-μm-thick sections using a diamond knife Diatome (Fig. 3k).
11. Check every three to five sections in a light microscope to obtain a central longitudinal section through the meristem (*see* **Note 20**).
12. Place central longitudinal sections and semi-thick sections in a drop of dH_2O on a poly-L-lysine microscope slide.
13. Let the sections dry at room temperature or heating plate (temperature not more than 38 °C), and do not cover (*see* **Note 21**).
14. Put slides into a slide storage box.

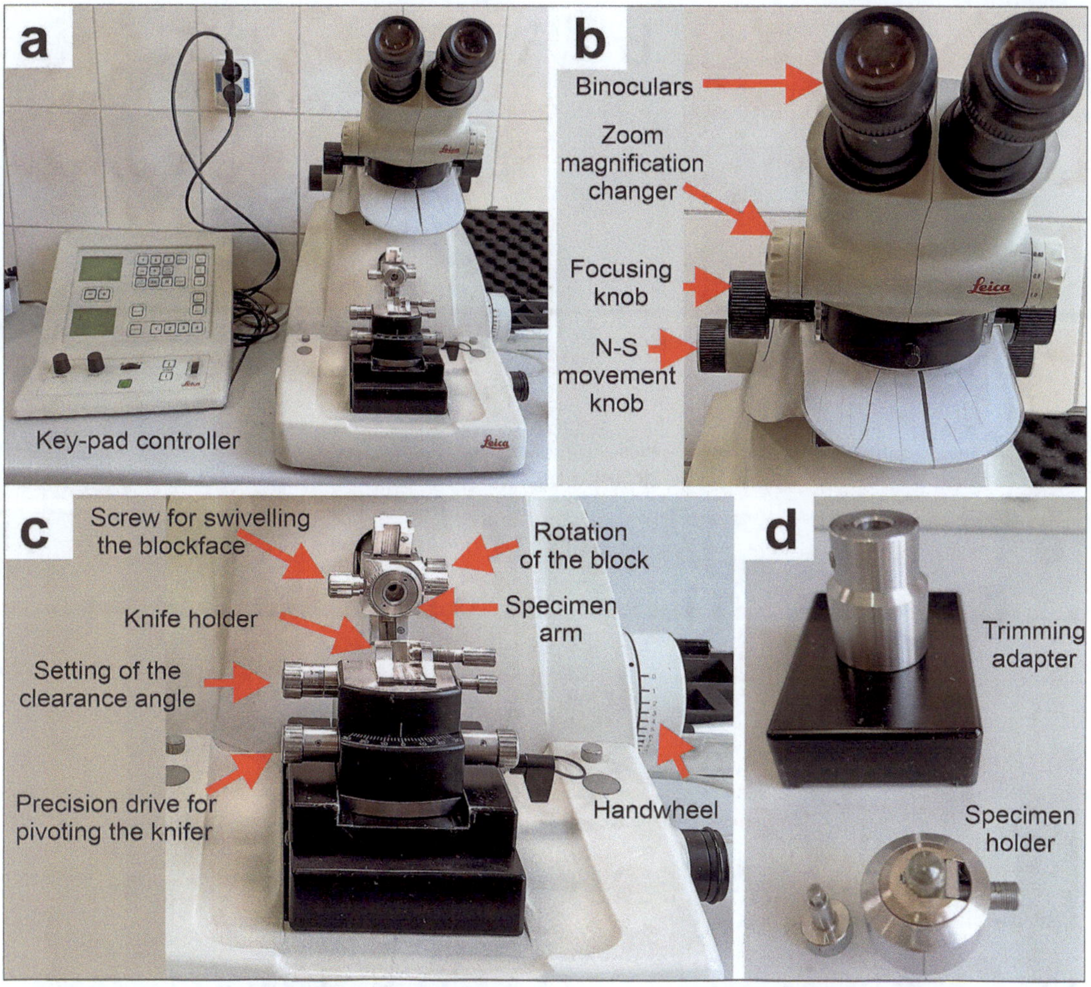

Fig. 2 Description of the ultramicrotome. (**a**) Microtome construction with a description of the elements. (**b**) The upper part of the microtome. (**c**) The lower part of the microtome with all elements of the specimen and knife holders. (**d**) Trimming adapter and specimen holder with the mounted sample

3.6 Immunostaining

Day 1

1. Take the prepared slides that have to be used for staining (Fig. 4a).
2. Outline the areas on the slide where the sections are located with a hydrophobic marker (PAP pen; Fig. 4b).
3. File the humid slide chamber with a wet paper towel (Fig. 4c–d).
4. Transfer the slides to a humid slide chamber.
5. Apply blocking buffer (2% BSA and 2% FCS on 0.1 M PBS) in the marked areas to cover the entire sections, and incubate for 30 min at room temperature (Fig. 4e; *see* **Note 22**).
6. Incubate slides with the primary antibody overnight at 4 °C (*see* **Note 23**).

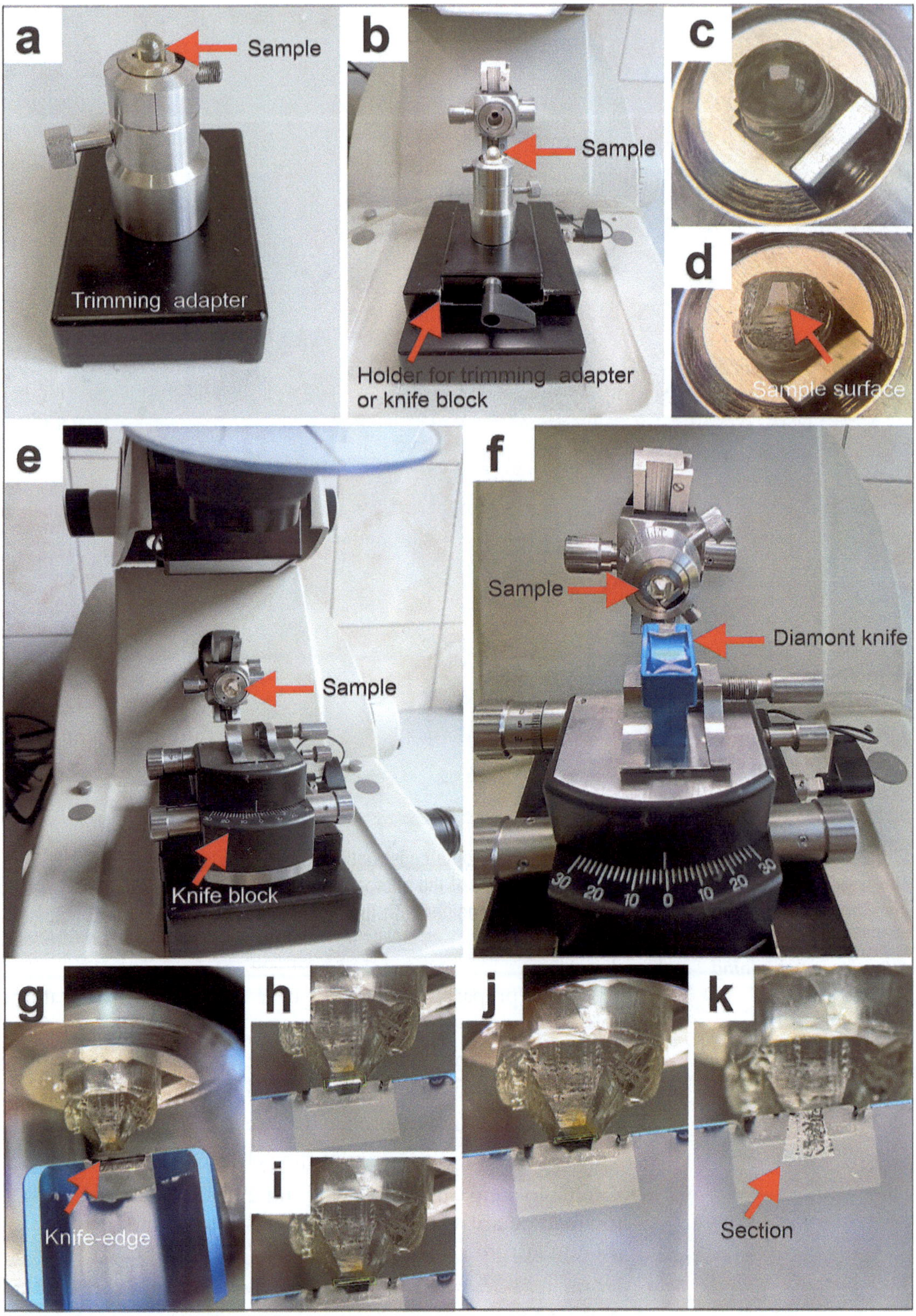

Fig. 3 Subsequent stages of sample preparation for cutting and correct positioning of the sample to the edge of the knife. (**a**) Sample in specimen holder ready for trimming. (**b**) Holder with a sample ready for trimming

Day 2

7. Wash slides in blocking buffer for 3 × 10 min.
8. Incubate the slides with secondary antibody 1.5–2 h at room temperature in the dark (*see* **Note 24**).
9. Wash slides in blocking buffer for 3 × 5 min.
10. Wash slides in 0.1 M PBS for 3 × 10 min.
11. Apply a FB28 for 5–10 min (*see* **Note 25**).
12. Wash slides in 0.1 M PBS for 3 × 10 min.

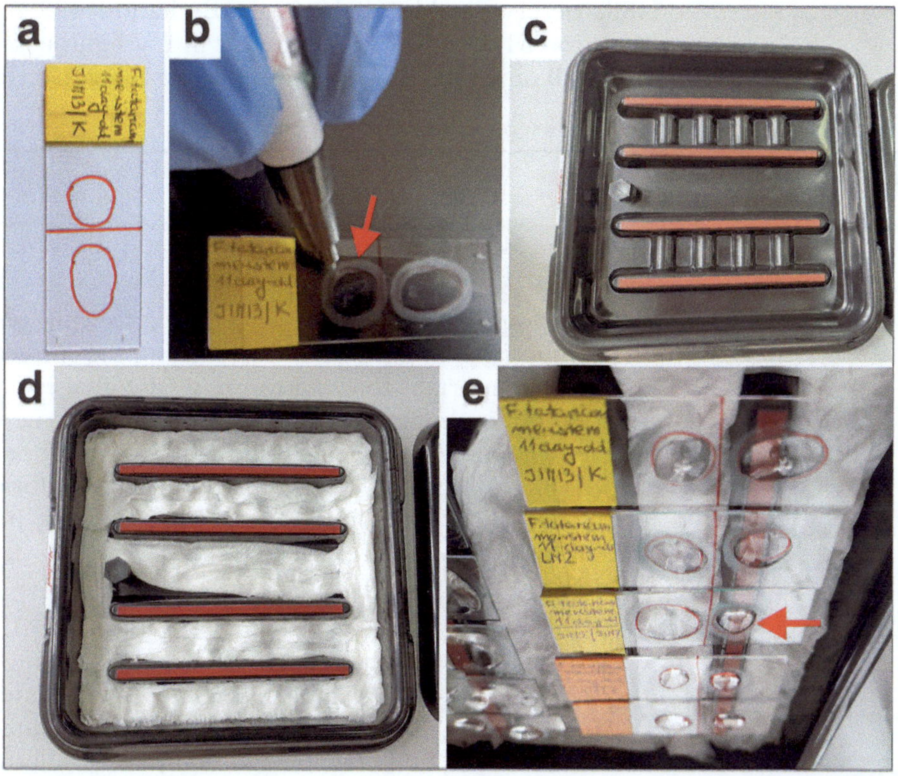

Fig. 4 Preparation of slides for immunostaining. (**a**) The microscope slide is divided into two parts, and the sections with the material are outlined with a circle. (**b**) Scraps with the material additionally outlined with a hydrophobic marker (PAP pen). (**c**) Humid slide chamber. (**d**) Humid slide chamber lined with a wet paper towel to retain moisture. (**e**) Drops of blocking buffer applied to a slide enclosed in the area outlined by PAP pen

Fig. 3 (continued) mounted in a holder of the microtome. (**c**) Sample before and (**d**) after trimming. (**e**) Sample in the specimen arm. (**f**) Position the sample in relation to the diamond knife. (**g–j**) Top view of the relative arrangement of the sample and the knife edge. The green rectangle shows the decrease in the distance of the sample from the knife edge by reducing the amount of light between the sample and the knife edge. (**k**) The section on the surface of the water in the knife bath

13. Wash slides in sterile ultrapure water for 3 × 10 min.
14. Remove the ultrapure water and gently dry the slides, and immediately after drying, apply two to three drops of Fluoromount (anti-fading medium), and cover with a coverslip (preferably 24 mm × 60 cm; *see* **Note 26**).
15. Transfer the slides into a slide storage box (preferably, the preparations should be in a horizontal position), and store them at 0–4 °C.

3.7 Observations

1. Observe in a fluorescent microscope at the indicated wavelength of light for the fluorochrome (for AlexaFluor® 488, observe in blue light: max. Excitation 493 nm, max. Emission 518 nm; for FB28 observe in UV light: max. Excitation 350 nm, max. Emission 432 nm; *see* **Note 27**) (Fig. 5).

4 Notes

1. Use 0.5 M NaOH to increase pH and 0.5 M HCl to decrease pH.
2. Fixation stops the living processes in the cells while preserving the composition of plant cells. Establishing the correct fixative mixture is important to avoid forming false images (artifacts). In addition, it should be remembered that the time between taking the sample and placing it in the fixative must be as short as possible. All fixation procedures should be performed in a well-functioning fume hood. Excess fixative can be portioned into falcons (15–20 mL) and stored at −20 °C for a few months.
3. 10% ethanol: add 10 mL of absolute ethanol to 90 mL distilled water (dH_2O); 30% ethanol: add 30 mL of absolute ethanol to 70 mL dH_2O; 50% ethanol: add 50 mL of absolute ethanol to 50 mL dH_2O; 70% ethanol: add 70 mL of absolute ethanol to 30 mL dH_2O; 90% ethanol: add 90 mL of absolute ethanol to 10 mL dH_2O.
4. Polypropylene or polypropylene flat-ended capsules can also be used.
5. One outline on slides is shown in Fig. 4e. The volume should be calculated to the number of slides analyzed. To prepare the second antibody stock solution, add an equal volume of glycerol (ACS grade or better) for a final concentration of 50%, and store it at −20 °C as a liquid.
6. Excess FB28 solution can be portioned into falcons (15–20 mL) and stored at −20 °C for a few months.

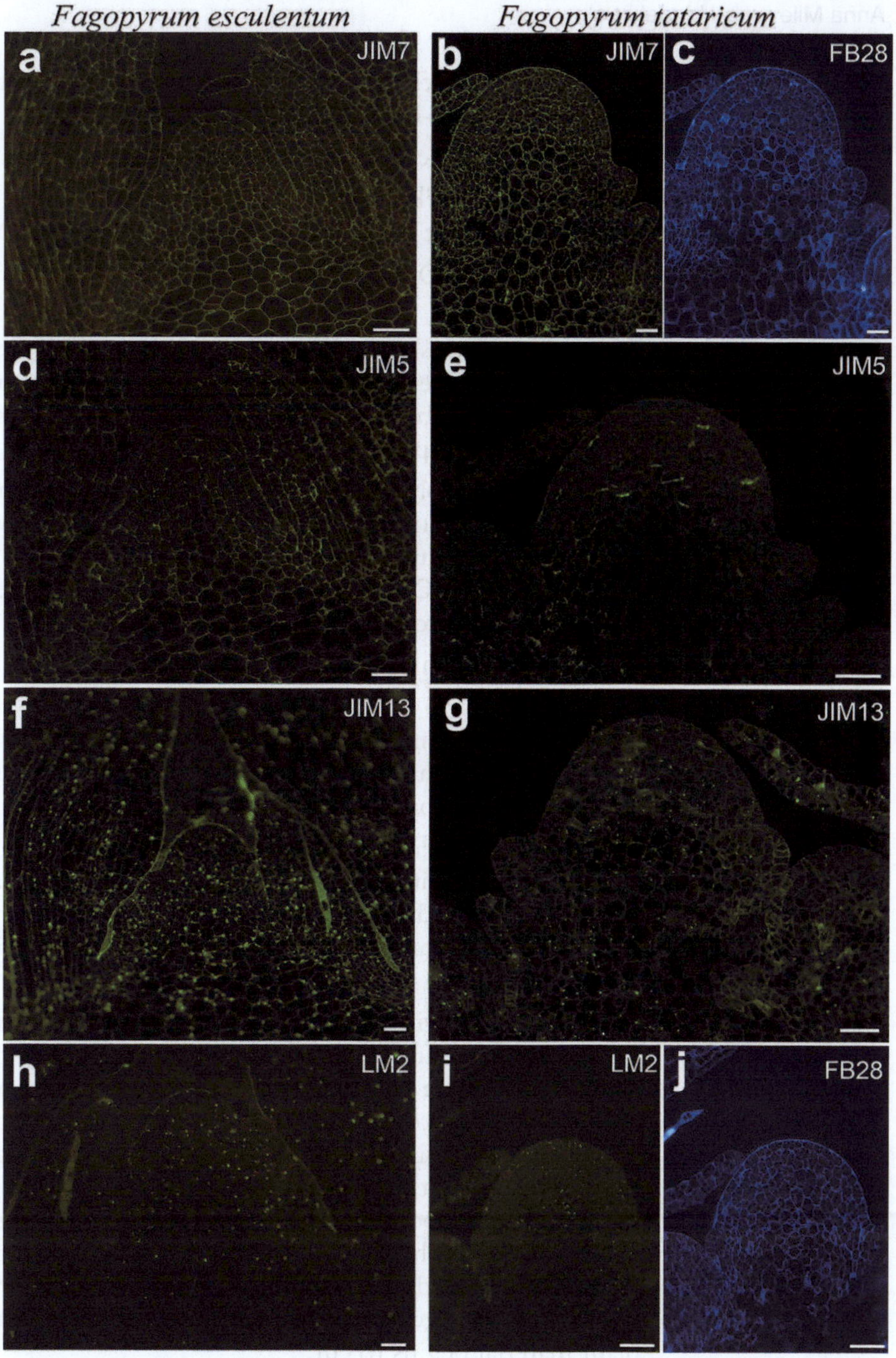

Fig. 5 Immunolocalization of pectic (JIM7, JIM5) and AGP (JIM13, LM2) epitopes in walls of shoot apical meristem cells of *Fagopyrum esculentum* and *Fagopyrum tataricum*. (**a**) Seven-day-old *F. esculentum* meristem showing JIM7 fluorescence. (**b**) Eleven-day-old *F. tataricum* meristem showing JIM7 fluorescence. (**c**) Eleven-day-old *F. tataricum* meristem showing FB28 fluorescence. (**d**) Seven-day-old *F. esculentum* meristem showing JIM5 fluorescence. (**e**) Eleven-day-old *F. tataricum* meristem showing JIM5 fluorescence. (**f**) Seven-day-old *F. esculentum* meristem showing JIM13 fluorescence. (**g**) Eleven-day-old *F. tataricum* meristem showing JIM13 fluorescence. (**h**) Seven-day-old *F. esculentum* meristem showing LM2 fluorescence. (**i**) Eleven-day-old *F. tataricum* meristem showing LM2 fluorescence. (**j**) Eleven-day-old *F. tataricum* meristem showing FB28 fluorescence. Scale bars – 50 μm

7. The humid slide chamber was a closed container with a wet paper towel at the bottom.
8. The volume of the fixative solution should be four times the volume of the sample.
9. If the PFA has not dissolved, add 100 uL of 1 M NaOH.
10. Use 0.5 M NaOH to increase pH and 0.5 M HCl to decrease pH.
11. Five to ten cycles, 10 min each; recommended pressure is 0.006 Pa until the samples sink completely to the bottom of the Eppendorf tube (Eppendorf lid should have holes to permit the air to escape).
12. Dehydration involves removing water from the sample and replacing it with organic solvents that are miscible with the embedding medium. Samples can be stored in 30% ethanol overnight at 4 °C. Samples in 70% ethanol can be stored overnight for up to several weeks at 4 °C.
13. It is very important to close the capsules well so there is no air access. Otherwise, LR white will not polymerize.
14. The sample sectioning is where the sample is cut off from the main source of material sections of a specific thickness. When cutting into sections, it is very important not to damage the material when cutting the sample.
15. It is important that the side of the block where the material is located is extended beyond the holder by about 10 mm. Moreover, the block must be stably and firmly fastened into the holder.
16. Place the holder with the block in the trimming adapter. Looking through the binocular, trim off excess resin on all four sides around the sample to create a trapezoidal shape. You must not cut off the excess resin with one cut; you can lose the material. Trim off excess resin by cutting thin pieces from the edge of the block until you get the right shape. Remember to change razor blades often as soon as they become dull. Then, cut off the excess resin from the top of the block. To do this, take the razor blade in both hands, holding its ends, and with decisive movements, gradually get rid of the excess resin from above until the plant material begins to cut.
17. Remember to tighten the screw so the holder is stable in the specimen arm.
18. Remember to tighten the screw so that the knife is stable.
19. You should see a shiny reflection on the block face – the narrower it is, the closer the knife is to the sample. Approach very slowly until the reflection disappears completely – then you are close enough to the sample to start slicing.

20. Transfer the section onto a drop of 0.05% TBO on a clean glass slide, and place it on a heating plate (40–50 °C). This staining is performed only for observing cross-section histology. During this staining, we can (1) judge if we are far from the central section and (2) properly correct the sample position if the obtained cross section is oblique.
21. In order to orientate where the sections are located, you can outline the bottom of the slide with a marker.
22. The whole procedure can be performed within 1 day, in which case the incubation time with the primary antibody is 1–1.5 h.
23. Remember to perform the negative controls – omit the primary antibody step (apply blocking buffer instead), and the rest of the procedure do the same. The absence of fluorescence is the expected result and indicates that the secondary antibody does not bind nonspecifically.
24. From this step, the procedure is performed in the dark!
25. For fluorescence detection of cellulose, visualize cell walls.
26. To extend the shelf life of the slides, the next day, outline the edges of the coverslip with nail polish.
27. Remember to take photos from the same area using two light filters (blue light and UV light) so that you can overlap the images from immunostaining (green fluorescence) and FB28 staining (blue fluorescence).

Acknowledgments

This work was supported by the National Science Centre Poland, project OPUS (2020/37/B/NZ9/01499).

References

1. Sala K, Malarz K, Barlow PW, Kurczynska E (2017) Distribution of some pectic and arabinogalactan protein epitopes during *Solanum lycopersicum* (L.) adventitious root development. BMC Plant Biol 17:25
2. Sala K, Godel-Jędrychowska K, Kurczyńska E (2022) Immunochemistry of cell wall – a tool for evaluation of the response of plants to changed habitat. In: Dyczko A, Jagodziński A, Woźniak G (eds) Green scenarios: mining industry responses to environmental challenges of the Anthropocene epoch: international mining forum 2021, 1st edn. CRC Press
3. Milewska-Hendel A, Sala K, Gepfert W, Kurczyńska E (2021) Gold nanoparticles-induced modifications in cell wall composition in barley roots. Cell 10:1965
4. Cosgrove DJ (2022) Building an extensible cell wall. Plant Physiol 189(3):1246–1277
5. Verhertbruggen Y, Marcus SE, Haeger A, Ordaz-Ortiz JJ, Knox JP (2009) An extended set of monoclonal antibodies to pectic homogalacturonan. Carbohydr Res 344:1858–1862
6. Willats WGT, Limberg G, Buchholt HC, van Alebeek G-J, Benen J, Christensen TMIE, Visser J, Voragen A, Mikkelsen JD, Knox JP (2000) Analysis of pectic epitopes recognised by hybridoma and phage display monoclonal antibodies using defined oligosaccharides, polysaccharides and enzymatic degradation. Carbohydr Res 327:309–320

7. Jones L, Seymour GB, Knox JP (1997) Localization of pectic galactan in tomato cell walls using a monoclonal antibody specific to (1[–>] 4)-[beta]-D-galactan. Plant Physiol 113:1405–1412
8. Willats WGT, Marcus SE, Knox JP (1998) Generation of a monoclonalantibody specific to (1-5)-a-L-arabinan. Carbohydr Res 308: 149–152
9. Moller I, Marcus SE, Haeger A, Verhertbruggen Y, Verhoef R, Schols H, Ulvskov P, Mikkelsen JD, Knox JP, Willats W (2008) High-throughput screening of monoclonal antibodies against plant cell wall glycans by hierarchical clustering of their carbohydrate microarray binding profiles. Glycoconj J 25: 37–48
10. Pennell RI, Janniche L, Kjellbom P, Scofield GN, Peart J, Roberts K (1991) Developmental regulation of a plasma-membrane arabinogalactan protein epitope in oilseed rape flowers. Plant Cell 3:1317–1326
11. Knox JP, Linstead PJ, Peart J, Cooper C, Roberts K (1991) Developmentally-regulated epitopes of cell surface arabinogalactan proteins and their relation to root tissue pattern formation. Plant J 1:317–326
12. Pennell RI, Knox JP, Scofield GN, Selvendran RR, Roberts K (1989) A family of abundant plasma membrane-associated glycopro-teins related to the arabinogalactan proteins is unique to flowering plants. J Cell Bioi 108: 1967–1977. https://doi.org/10.1083/jcb.108.5.1967

Chapter 7

Immunodetection of Cell Wall Components in Studies on Cell Wall Rebuilding in *Fagopyrum esculentum* and *Fagopyrum tataricum*

Anna Milewska-Hendel, Katarzyna Sala-Cholewa, and Reneé Pérez-Pérez

Abstract

Immunocytochemical studies of the cell wall are used to visualize specific epitopes of pectins, arabinogalactan proteins, hemicelluloses, extensins, and other wall components using specific primary antibodies. This reaction, combined with calcofluor staining, allows to comprehend how the cell wall is rebuilt during the protoplast culture. In this protocol, the method of immunostaining using antibodies against cell wall components based on *Fagopyrum esculentum* and *Fagopyrum tataricum* protoplasts is described.

Key words Arabinogalactan proteins, Cell wall, *Fagopyrum*, Pectins, Protoplasts

1 Introduction

The cell wall is a dynamic and complex part of plant cells, which, depending on its chemical composition, is able to perform various functions. The scaffolding of the cell wall is made of cellulose microfibrils that are immersed in an amorphous matrix that consists of pectins, hemicelluloses, structural proteins (e.g., arabinogalactan or extensin), enzyme proteins, or phenolic compounds [1, 2].

One of the methods allowing for qualitative analysis of the occurrence of individual components of the cell wall is immunocytochemistry. Indirect reactions are most often used, where two-step reactions are applied. The first step uses a primary antibody directed against selected epitopes of cell wall components. In the subsequent reaction, secondary antibodies conjugated with a fluorochrome are employed [3]. A large resource of primary antibodies can be used in the research, namely, those raised against pectin, arabinogalactan proteins (AGPs), extensins, or hemicelluloses epitopes. Immunostaining allows for visualization and determination of the composition of the cell wall in various types of cells, at

Alexander Betekhtin and Artur Pinski (eds.), *Buckwheat: Methods and Protocols*, Methods in Molecular Biology, vol. 2791, https://doi.org/10.1007/978-1-0716-3794-4_7,

different stages of development in response to changing environmental conditions. Thus, these studies also make it possible to understand the function of individual cell wall components [4].

Protoplasts are cells where the cell wall has been enzymatically or mechanically removed. Protoplasts are an excellent system for genetic manipulation technologies as they, in theory, are totipotent [5, 6]. It means that they have the ability to differentiate, proliferate, or regenerate into various organs or whole plants [7]. However, the cell wall must be rebuilt for these processes to be successfully carried out. Immunocytochemical methods with the use of specific antibodies detecting cell wall components are the best tools that can be implemented for this research.

2 Materials

2.1 Protoplast Staining

1. Washing solution (W5; 154 mM NaCl, 125 mM $CaCl_2{\cdot}2H_2O$, 5 mM KCl, 5 mM glucose). For preparation, dissolve the following in 200 mL distilled water: 2.250 g of NaCl, 4.594 g of $CaCl_2{\cdot}2H_2O$, 0.093 g of KCl, and 0.225 g of glucose. Adjust the pH to 5.8 and the volume to 250 mL with distilled water. Sterilize by autoclaving (0.1 MPa, 121 °C), and store at room temperature.
2. Isolated protoplasts in W5 solution.
3. Fuchs-Rosenthal hemocytometer.
4. Automatic pipette with sterile tips.
5. Ibidi µ-Dish (Ø 35 mm, low).
6. Vacuum filter.
7. 0.22 µm filters.
8. 0.01% propidium iodide (PI) solution in W5: dissolve 0.01 g of PI in 100 mL of W5, sterilize by filtration under the fume hood and store at 4 °C, and protect from light.
9. 0.01% fluorescent brightener 28 (FB28) solution in W5: dissolve 0.01 g of FB28 in 100 mL of W5, sterilize by filtration under the fume hood and store at 4 °C, protect from light.
10. Inverted OLYMPUS IX81-ZDC microscope with fluorescence and bright-field image acquisition (*see* **Note 1**).

2.2 Protoplast Selection

1. Micromanipulator Eppendorf TransferMan® 4r.
2. Stripper tips MXL3-75 (ORIGIO, Inc.).
3. 1.5 mL Eppendorf tubes.
4. Vacuum filter.
5. 0.22 µm filters.

6. Basal medium (BM) for protoplast cultivation. Dissolve the following in 800 mL distilled water: 3.623 g of KM macro-microelements and organic acids, 0.10 g of myoinositol, 10 mg of thiamine, 1 mg of pyridoxine, 1 mg of nicotinic acid, 0.25 g of N-Z-amineA, 1 mg of 6-benzylaminopurine (BAP), 2 mg of 1-naphthalenacetic acid (NAA), and 74 g of glucose. Adjust the pH to 5.6, and the volume to 1 L. Sterilize by filtration under the fume hood and store at 4 °C.
7. Ice.

2.3 Protoplast Culture

1. Microwave.
2. Vacuum filter.
3. 0.22 μm filters.
4. 1.2% low melting point agarose (LMPA): weigh 1.2 g of Sea-Plaque agarose, and dissolve in 100 mL of BM using a microwave. Sterilize by filtration under the fume hood, and store at room temperature.
5. 100 μM working solution of phytosulfokine (PSK): weight 0.1 g of PSK and dissolve in 1 mL of filter-sterilized, distilled water under the fume hood. Store at −20 °C.
6. 8 mg/L working putrescine (PUT) solution: dissolve 2 mg of PUT in 250 mL of distilled water. Sterilize by filtering and store at -20 °C (*see* **Note 2**).
7. 0.25 mg/L working solution of chloropyridin phenylurea (CPPU): dissolve 0.25 mg of CPPU in 1 L of distilled water. Sterilize by filtering and store at -20 °C (*see* **Note 3**).
8. 500 mg/mL Timentin antibiotic solution: dissolve 12.5 g of Timentin in 25 mL of distilled water. Sterilize by filtering and store at −20 °C (*see* **Note 4**).
9. Polystyrene Petri dishes (Ø 6 cm × 1.2 cm).
10. Parafilm.
11. Laboratory incubator (25 °C).

2.4 Fixation

1. 1× phosphate buffered saline (PBS). Dissolve 8 g NaCl, 0.2 g KCl, 0.24 g KH_2PO_4, and 1.44 g Na_2HPO_4 in 1 L of distilled water. Adjust the pH to 7.2 (use 0.5 M NaOH to increase pH and 0.5 M HCl to decrease pH).
2. Glass beaker.
3. 2 mL Eppendorf tubes.
4. Scalpel and/or forceps.
5. Plastic fine tip pipettes (5 mL).
6. Plastic pipettes (5 mL).
7. pH meter.

8. Heating plate
9. Aluminum foil
10. 3% paraformaldehyde (PFA) and 1% glutaraldehyde (GA) fixative. To prepare a 50 mL fixative solution, use 1.5 g PFA and 2 mL of 25% GA (or 1 mL of 50% GA) in 1× PBS (v/v) (*see* **Note 5**).

    Preparation of 50 mL fixative

    A. Add 40 mL of 1× PBS to a glass beaker (greater than 50 mL and less than 100 mL).

    B. Weigh out 1.5 g of PFA and pour it into the beaker with 1× PBS.

    C. Cover the beaker with aluminum foil and heat on the heating plate set for approximately 50 °C for 1–1.5 h (if there is still undissolved PFA at the bottom of the beaker, it is recommended to add 100 μL of 1 M NaOH).

    D. Cool the fixative solution and filter through a filter paper.

    E. Add 2 mL of 25% GA or 1 mL of 50% GA.

    F. Add PBS to a volume of 50 mL.

    G. Adjust the pH to 7.2.

2.5 Immunostaining

1. Blocking buffer consisting of 2% BSA (bovine serum albumin) and 2% FCS (fetal calf serum) in 1× PBS buffer (add 1 g of BSA and 1 mL of FCS to 49 mL of 1× PBS) (*see* **Note 6**).
2. Primary antibodies working solutions, dilution 1:20 in blocking buffer (1 μL antibody per 19 μL blocking buffer).
3. Secondary antibody goat anti-rat IgG H + L conjugated with fluorochrome AlexaFluor®488, dilution 1:100 in blocking buffer (1 μL antibody per 99 μL blocking buffer).
4. 0.01% solution of Fluorescence Brightener 28 (FB28, calcofluor) in 1× PBS: dissolve 0.01 g of FB28 in 100 mL of PBS, shake until dissolved, and subsequently filter through a 0.22 μm syringe filter (*see* **Note 7**).
5. Ultrapure water.

2.6 Microscopical Observations

1. Olympus FV1000 confocal system, equipped with an Olympus IX81 inverted microscope, an ×60 Plan Apo oil-immersion objective lens (NA 1.35), a 50 mW diode laser (excitation 405 nm), a 100 mW multiline (excitations 457/488/515 nm) argon ion laser, two separate photomultipliers, and epifluorescence lamp.
2. Microscopic slides and coverslips.
3. Immersion oil.
4. Plastic fine tip pipettes (5 mL).

3 Methods

3.1 Protoplast Staining

1. Count the protoplasts using a hemocytometer, and adjust the concentration to 800,000 cells/mL, adding W5 solution if necessary (*see* **Note 8**).
2. Dissolve 0.5 mL of the protoplast solution in 9.5 mL of W5, and keep the rest in the fridge (*see* **Note 9**).
3. Pipette 0.8 mL of the obtained protoplast solution into the Ibidi dish well, and add 100 μL of PI (to select viable cells that do not exhibit fluorescence signal) and 100 μL of FB28 (to select cells without cell walls that do not exhibit fluorescence signal).
4. Close the dish, place it on the inverted microscope, and incubate in the dark for 15 min (*see* **Note 10**).

3.2 Protoplast Selection

1. Carefully open the dish and use a micromanipulator (Fig. 1a, b) to focus the needle tip (stripper tip) in the same plane as the protoplasts. Use the 20× objective and a bright field.
2. With the use of fluorescence filters, pick up the protoplasts that do not present FB28 or PI signal.
3. Discharge the contents of the needle into a sterile 1.5 mL Eppendorf tube that contains 5 μL of BM medium on the bottom. Keep it closed on ice.
4. Load the needle again, and repeat the previous steps in the same Eppendorf tube until around 200 cells are collected (*see* **Note 11**).

3.3 Protoplast Culture

1. Add 6 μL of warm LMPA to the Eppendorf tube with selected protoplasts, and mix it carefully. Pipette the total volume (~11 μL) into a Petri dish, forming a small drop (*see* **Note 12**).
2. For the nurse culture, mix 1 mL of the 800,000 cells/mL protoplast suspension with 1 mL of warm LMPA in a new tube. Form three drops of 50 μL around the drops containing the selected protoplasts in the Petri dish (Fig. 2a). Wait a few minutes for the drops to solidify (Fig. 2b).
3. Add 4 mL of BM medium supplemented with 4 μL PSK, 4 μL PUT, 4 μL CPPU, and 2.4 μL Timentin.
4. Seal the Petri dishes with parafilm, and incubate them in the dark at 25 °C for 4 h, 12 h, 24 h, 48 h, and 72 h (*see* **Notes 13** and **14**).

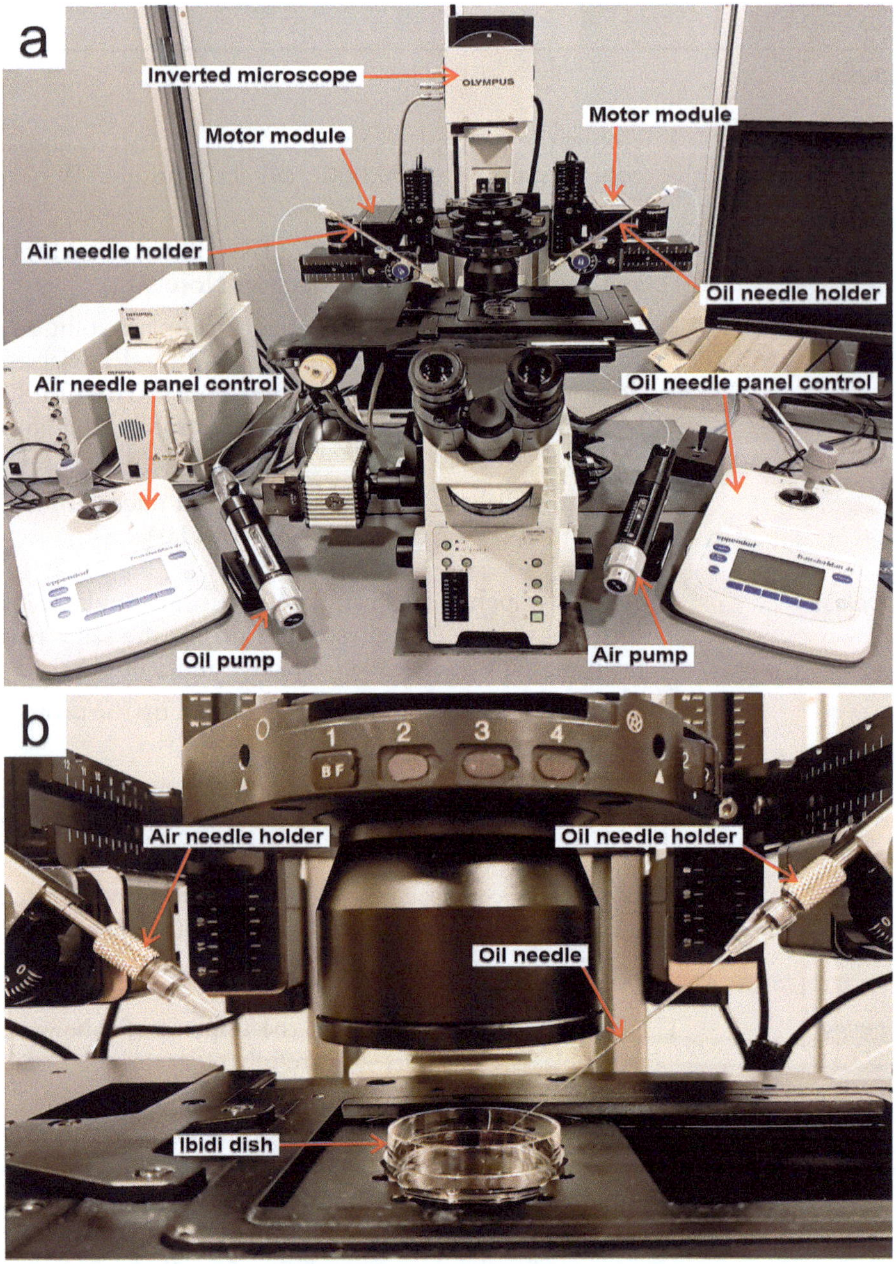

Fig. 1 Single-cell micromanipulator (Eppendorf TransferMan® 4r) installed on an inverted microscope (OLYMPUS IX81-ZDC) (**a**). Needle connected to the oil pump (**b**)

3.4 Fixation

1. After the established incubation time, prepare 2 mL round-bottom Eppendorf tubes with the addition of 1 mL of 3% PFA and 1% GA fixative.
2. Remove the BM medium from a Petri dish with a Pasteur pipette, and transfer the selected protoplast drops (no more than ten drops per Eppendorf tube) into Eppendorf tubes with a fixative using a scalpel.

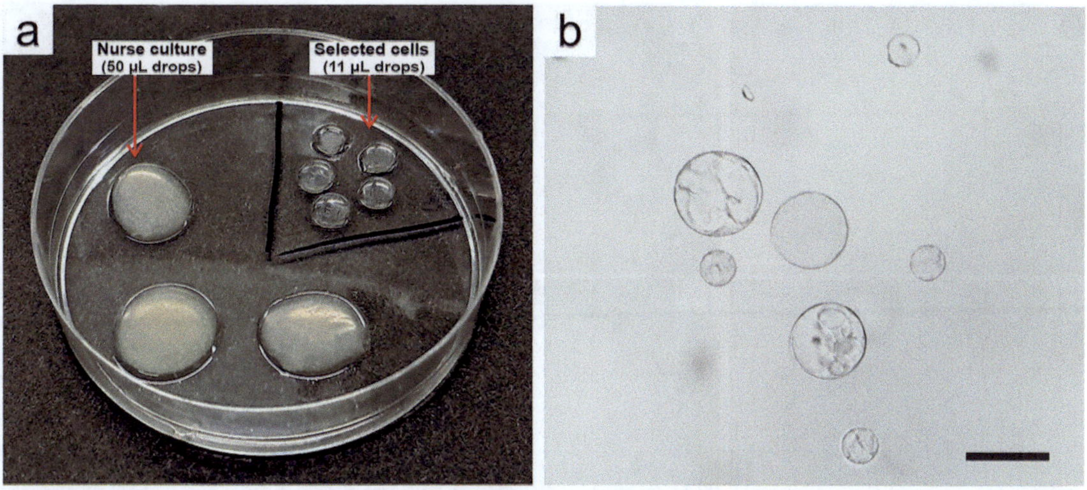

Fig. 2 Polystyrene Petri dish (Ø6 cm × 1.2 cm) with three drops of 50 μL of nurse culture (800,000 cells ml^{-1}) and five ~11 μL drops of selected *F. tataricum* (200 cells per drop) (**a**). Bright-field view of selected *F. tataricum* protoplasts embedded in LMPA (time 0 h) (**b**). Scale bar: 50 μm

3. If needed, add fixative to reach 2 mL (*see* **Note 15**).
4. Incubate the samples overnight at 4 °C.
5. Next day, remove the fixative and wash the drops with PBS three times for 10 min each (*see* **Note 16**).

3.5 Immunostaining

1. Prepare the appropriate number of 2 mL Eppendorf tubes according to the number of antibodies you want to apply and one Eppendorf tube for the negative control (*see* **Note 17**).
2. Apply 1.5 mL of blocking buffer to each Eppendorf tube.
3. Transfer three to five agarose drops with protoplasts to each Eppendorf tube, and incubate in a blocking buffer for 30 min at room temperature.
4. Remove the blocking buffer and add 250 μL of each primary antibody. Incubate at room temperature for 1.5–2 h (*see* **Notes 18** and **19**).
5. Remove the primary antibody solution, and wash samples in blocking buffer three times for 10 min.
6. Incubate the samples with 250 μL of secondary antibody for 1.5–2 h at room temperature in the dark (*see* **Note 20**).
7. Remove the secondary antibody solution, and wash samples in 1× PBS three times for 10 min.
8. Apply a FB28 for 15 min (*see* **Note 21**).
9. Remove FB28 and wash samples in 1× PBS three times for 10 min.
10. Wash samples in sterile ultrapure water three times for 10 min.
11. Keep stained samples in sterile ultrapure water at a temperature of 0–4 °C in the dark (*see* **Note 22**).

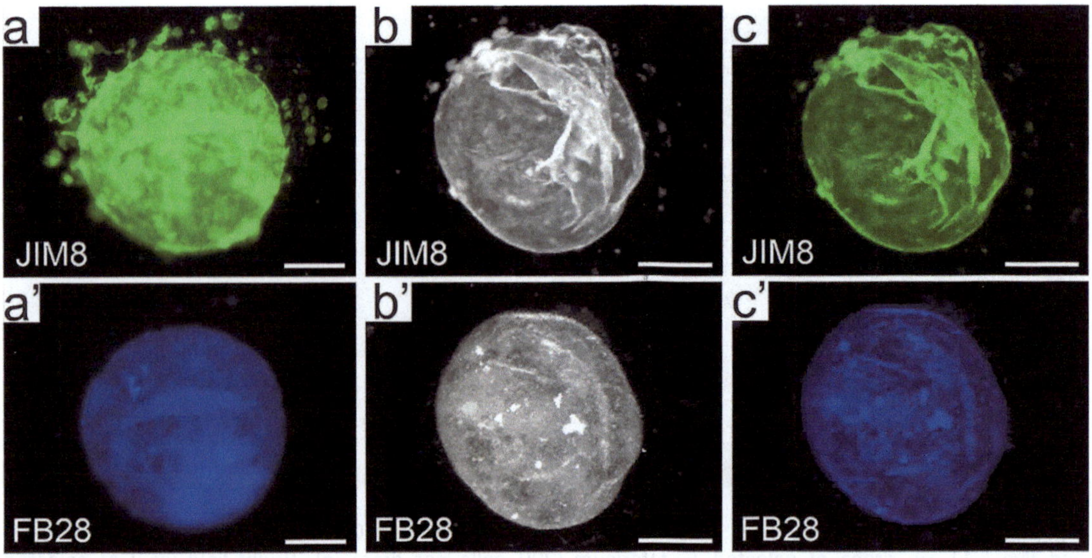

Fig. 3 Epifluorescence (**a**, **a′**) and confocal (**b**, **b′**, **c**, **c′**) images of stained protoplasts (primary antibody for blue light channel: JIM8; histochemical staining for UV channel: FB28). Images from epifluorescence microscope viewed in blue light (**a**) and UV channel (**a′**); confocal microscope: blue light laser – 488 nm wavelength (**b**), UV laser – 405 nm (**b′**), saved in grayscale; blue light laser – 488 nm wavelength (**c**), UV laser – 405 nm (**c′**), saved in colored version with the use of LUT tool. Scale bar: 10 μm

3.6 Microscopical Observations

1. Take out an agarose drop of stained protoplasts from the Eppendorf tube and put it on a microscopic slide; close gently with a coverslip.
2. With the use of epifluorescence preview, find the protoplasts.
3. Apply immersion oil and begin scanning in two channels: Alexa 488 fluorochrome (for excitation, select the 488 nm laser) and FB28 (for excitation, use 405 nm laser) (Fig. 3) (*see* **Note 23**).

4 Notes

1. To excite FB28 and PI fluorescence, UV and green light spectra are essential.
2. The working solution can be divided into smaller aliquots.
3. The working solution can be divided into smaller aliquots.
4. The solution can be divided into smaller aliquots.
5. Fixation kills the cells while preserving their composition. Fixed samples will not decompose; however, a proper fixative mixture should be used to avoid the formation of artifacts in microscope images. The collected material must be put into fixative immediately to avoid sample degradation or any

physiological changes. To avoid inhaling, all fixation steps must be performed in a well-functioning fume hood. The excess fixative can be portioned into falcons (15–20 mL) and stored at -20 °C until the change of color (up to a few months).

6. Wait until BSA dissolves; do not shake.
7. Excess calcofluor stain can be portioned into plastic falcons (exemplary volume 15 mL) and stored at −20 °C for a few months.
8. The counting procedure is described in another chapter.
9. Decreasing the concentration of protoplasts guarantees an adequate separation between them, facilitating the subsequent selection of individual cells using the micromanipulator.
10. This step stabilizes the protoplasts on the well bottom.
11. Do not keep the Eppendorf tubes on ice for an extended period of time. Culture the protoplasts immediately after collecting 200 cells.
12. Up to five drops can be placed in the same Petri dish.
13. Dedicate one to three Petri dishes for each time point. Prepare many drops with ~200 protoplasts for technical replicates for each time point.
14. For 0 h time point, drops with protoplasts should be immediately fixed after selection.
15. The amount of fixative should be four times the volume of the sample.
16. It is preferable to start the immunostaining reaction immediately after the fixing procedure; however, fixed protoplasts can be kept in PBS at 4 °C until the antibody treatment for a short period of time.
17. For protoplast immunostaining, cells are stained in prepared and fixed agarose drops. They do not require dehydration and embedding in the medium. The purpose is the observation of whole cells, not sections.
18. For example, to prepare 250 μL of X primary antibody solution, take 12.5 μL of antibody and add 237.5 μL of blocking buffer (dilution 1:20).
19. For the negative control, leave the samples in a blocking buffer instead of applying a primary antibody. A negative control is carried out to confirm the specificity of the secondary antibody. Therefore, no fluorescence signal is expected.
20. Please note that once the secondary antibody with fluorochrome is added, you must work in the dark. The fluorochrome is photosensitive, and prolonged exposure to light can cause a false absence of the fluorescence signal.

21. Calcofluor staining is crucial in these studies. It allows us to estimate when the main component of the cell wall, cellulose, appears.
22. Make observations as soon as possible, up to a maximum of 2 weeks after staining.
23. Images of stained protoplasts can also be successfully obtained in an epifluorescence microscope; however, confocal images are more detailed, and optical sections can be made.

Acknowledgments

This work was supported by the National Science Centre Poland, project SONATA BIS 10 grant number 2020/38/E/NZ9/00033.

References

1. Cosgrove DJ (2005) Growth of the plant cell wall. Nat Rev Mol Cell Biol 6(11):850–861
2. Cosgrove DJ (2016) Plant cell wall extensibility: connecting plant cell growth with cell wall structure, mechanics, and the action of wall-modifying enzymes. J Exp Bot 67(2):463–476
3. Willats WG, Knox JP (2003) Molecules in context: probes for cell wall analysis, vol 8. Blackwell Publishing/CRC Press, Oxford/Boca Raton, pp 92–110
4. Knox JP (2008) Revealing the structural and functional diversity of plant cell walls. Curr Opin Plant Biol 11(3):308–313
5. Davey MR, Anthony P, Power JB, Lowe KC (2005) Plant protoplast technology: current status. Acta Physiol Plant 27:117–130
6. Eeckhaut T, Lakshmanan PS, Deryckere D, Van Bockstaele E, Van Huylenbroeck J (2013) Progress in plant protoplast research. Planta 238: 991–1003
7. Verma N, Bansal MC, Kumar V (2008) Protoplast fusion technology and its biotechnological applications. Chem Eng Trans 14:113–120

Chapter 8

Visualization of *Fagopyrum esculentum* and *Fagopyrum tataricum* Chromosomes and Micronuclei

Reneé Pérez-Pérez and Jolanta Kwasniewska

Abstract

This chapter presents the squash chromosome preparation technique for *Fagopyrum esculentum* and *F. tataricum*, using the root tips as the source of the material. Using an optimized version of this method, the chromosomes are free of cytoplasmic debris and are spread evenly on the glass slide. What comes of it is the possibility to make observations of the chromosome number and structure at the metaphase stage. This technique's modified version allows micronuclei analysis in interphase cells of buckwheats.

Key words Buckwheat, Chromosomes, Enzyme cocktail, *Fagopyrum*, Micronuclei, Mitosis, Squashed slides

1 Introduction

Chromosome-based technology has contributed significantly to bringing together general biology, biotechnology, and mutagenesis [1]. A breakthrough in chromosome preparations started with the development of the squash technique, which replaced the time-consuming sectioning procedures. Some other associate procedures, such as the pretreatment method, were integrated into the squash chromosome preparation technique. Many techniques, including chemical ones, are used for material pretreatment in buckwheat [2]. All the procedures must be optimized to the species and type of material used. A squash chromosome technique should allow for good spreading of the chromosomes during squashing and show details of chromosome morphology at the same time [2–4]. It is common knowledge that some plant species are among the most difficult regarding the squash method. The knowledge of the nuclear genome stability, described as chromosome numbers and structure, as well as the presence of the micronuclei in buckwheat, is crucial in improving projects on this crop [5].

Alexander Betekhtin and Artur Pinski (eds.), *Buckwheat: Methods and Protocols*, Methods in Molecular Biology, vol. 2791,
https://doi.org/10.1007/978-1-0716-3794-4_8,

2 Materials

2.1 Plant Material, Disposables, and Equipment

1. Roots from germinated seeds or plants of *F. esculentum* and *F. tataricum*.
2. Glass maceration containers.
3. Fine tweezers.
4. Needles.
5. Scalpel.
6. Blotting paper.
7. Micropipettes.
8. Eppendorf tubes, 1.5 mL.
9. Dry ice.
10. Glass slides, stored in 100% ethanol and dried before use.
11. Coverslips (22 × 22 mm, 24 × 24 mm).
12. Ø14 cm Petri dishes.
13. Incubator set at 37 °C.
14. Laboratory scale.

2.2 Solutions

Prepare all solutions using distilled water. Prepare all reagents at room temperature. Store all reagents at 4 °C unless indicated otherwise.

1. Fixative buffer: ethanol-acetic acid (EtOH-AC). Mix 3:1 ethanol or methanol with acetic acid. Measure and mix well just before use.
2. 10 mM citrate buffer: dissolve 0.84 g of citric acid monohydrate ($C_6H_8O_7.H_2O$) in 40 mL of distilled water (0.1 M) and 1.76 g of trisodium citrate dihydrate ($C_6H_5O_7Na_3.2H_2O$) in 60 ml of distilled water (0.1 M). Mix both solutions and dilute them 10×. Keep at −20 °C. Prewarm before use.
3. Enzyme cocktail: 2% w/v of cellulase Onozuka RS and 20% w/v of pectinase in 10 mM citrate buffer. Divide into 1 mL aliquots in Eppendorf tubes and store at −20 °C. Each aliquot can be reused three to five times for the same material.
4. 45% acetic acid.
5. DAPI staining solutions: mix DAPI and VECTASHIELD® in 1:10. Protect it from the light and store it at 4 °C.

3 Methods

Carry out all steps at room temperature unless otherwise specified.

3.1 Seed Germination

F. esculentum: keep the seeds in tap water overnight to soften the seed coat. Remove the coat, and place the seeds on filter paper moistened with tap water in a Petri dish. Incubate at 22 °C in the dark until the roots reach a length of approximately 1–2 cm.

F. tataricum: place the seeds on filter paper moistened with tap water in a Petri dish. Incubate at 22 °C in the dark until the roots reach a length of approximately 1–2 cm.

3.2 Pretreatment and Slide Preparation

1. Cut the 2 cm root fragments from seedlings or plants. Place them in 5 mL Eppendorf tubes with cold tap water for the low-temperature pretreatment stage to accumulate the condensed chromosomes. Incubate for 24 h on ice, leaving the Eppendorf tubes open (*see* **Note 1**).
2. Remove the water and fix immediately in freshly prepared EtOH-AC solution for at least 2 h (Fig. 1). The material can be stored in the fixative at −20 °C until required (*see* **Note 2**) (Fig. 1).
3. Shorten the fixed roots to a length of about 0.5 cm, and place them in a glass maceration container using a fine tweezer. Wash the roots in 10 mM citrate buffer 3 × 5 min to remove the fixative (Fig. 2).
4. Macerate the root tips by dipping them into the enzyme cocktail, previously prewarmed to 37 °C in a glass maceration container by the incubation for 1.5 h at 37 °C for chromosomes slides (*see* **Notes 3** and **4**) and 1.1 h at 37 °C for nuclei and micronuclei slides (*see* **Note 5**).

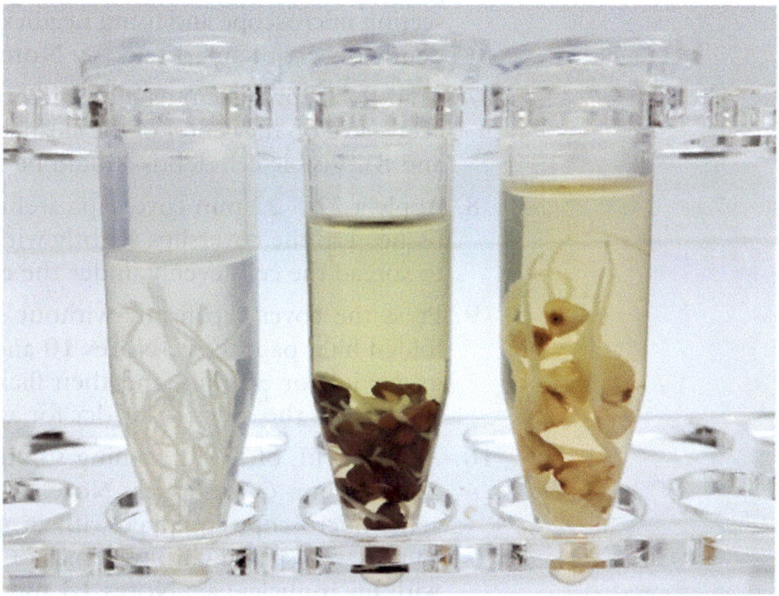

Fig. 1 Roots and seedlings fixed in EtOH-AC solution. From left to right: roots from *F. esculentum* cultivated in rooting medium; seedlings of *F. tataricum* and *F. esculentum*

Fig. 2 *F. esculentum* (**a**) and *F. tataricum* (**b**) roots in citrate buffer. *F. esculentum* roots in enzyme maceration solution (**c**) and root tips in 45% acetic acid (**d**)

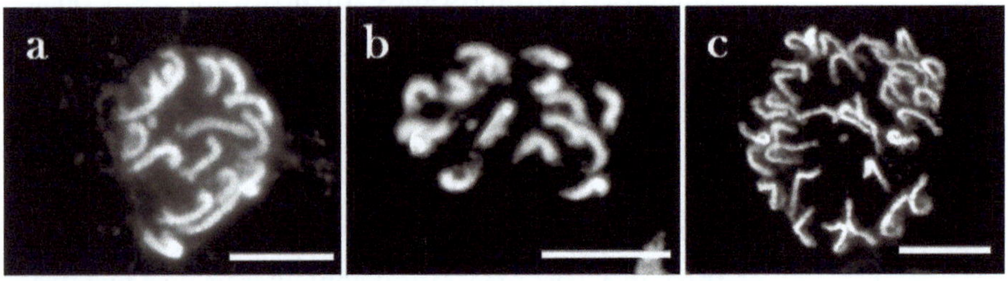

Fig. 3 Examples of *F. esculentum* (**a**) and *F. tataricum* (**b**) metaphase chromosome karyotype (2n = 16). (**c**) Example of tetraploid *F. esculentum* homokaryon (4n = 32). Scale bar: 10 μm

5. Remove carefully the enzyme solution using micropipettes, and add 1 mL of 10 mM citrate buffer. Leave the roots in 10 mM citrate buffer for 30 min.
6. Using a 1 mL micropipette, transfer the root tip into a new glass maceration container with 45% acetic acid. Under a dissecting microscope and using needles, remove the root cap and gently dissect the meristem (*see* **Note 6**).
7. Transfer the meristem using a micropipette to the 4 μL of 45% acetic acid on the center of the clean, dry glass slide (*see* **Notes 7** and **8**). Visible cell debris should be pinched away.
8. Apply a 22 × 22 mm coverslip carefully onto the meristematic tissue. Tap the coverslips slightly with the needle or toothpick to spread the cells evenly under the cover slip (*see* **Note 9**).
9. Press the cover slip firmly without sliding it. Do it between folded filter papers (see **Notes 10** and **11**). Incubate the slides in dry ice for 30 min, and then flick off the coverslip with a scalpel. Let the slides to air-dry for at least 30 min.
10. Apply 10 μL of DAPI staining solution to each slide. Apply a 24 × 24 mm coverslip (*see* **Note 12**). Analyze with a fluorescence microscope equipped with an appropriate filter for DAPI to find well-spread chromosomes or well-preserved cell nuclei with micronuclei (see **Notes 13** and **14**) (Figs. 3 and 4).
11. The slides can be stored at 4 °C for up to 3 months.

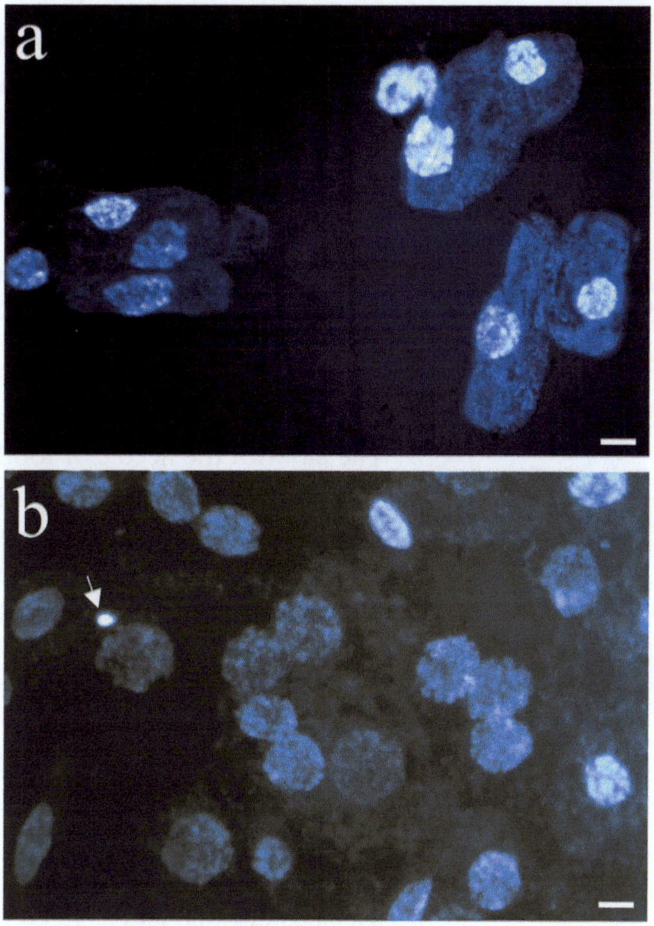

Fig. 4 Examples of *F. esculentum* control, without micronuclei (**a**) and mutagen-treated (**b**) nuclei; arrowhead points to micronucleus. Scale bar: 10 μm

4 Notes

1. To analyze micronuclei during interphase as a clastogenic effect of mutagens, the pretreatment stage is not used, and the next step is immediately passed.
2. For long-term storage (longer than 3 months) of root tips, replace the fixative solution with 70% ethanol and store them at 4 °C.
3. Described maceration conditions are suitable for uniform digestion of the cell wall of buckwheat cells. Although the longer you use the enzyme, its strength becomes weaker, and maceration time should be extended accordingly, or a new enzyme should be prepared. After use, centrifuge in a micro-centrifuge, transfer the supernatant to a new tube, mark for reuse, and freeze.

4. 1 mL of the enzyme cocktail is enough for the maceration of around five to ten root tips 0.5 cm long.
5. Shortening the time of enzymatic maceration allows the preservation of the cytoplasm and thus enables the analysis of the presence of micronuclei in a particular cell.
6. Avoid contamination with root caps cells, as they hinder the spreading of meristematic cells.
7. Due to the large size of the buckwheat meristems, one meristem is used to make one chromosome or micronuclei preparation.
8. This step must be done quickly because the roots are overexposed and become very soft.
9. The tap with the needle or toothpick can be unnecessary.
10. It is unnecessary to press hard if the meristematic cells are adequately macerated and dissected. Excess pressure results in the breakage of chromosomes and the far-spreading of chromosomes from one metaphase plate.
11. To analyze the micronuclei, you need to squash the material very gently not to damage the micronuclei, which are much smaller than cell nuclei.
12. 22 × 22 mm glass coverslips are used. Additionally, spread metaphases are often near the periphery of the coverslip, and slightly larger coverslips are then used for visualization (24 × 24 mm).
13. A normal diploid cell of *F. esculentum* and *F. tataricum* has 16 chromosomes ($2n = 16$).
14. If too much cytoplasm is present, alcohol-acetic acid fixative (3: 1) at −20 °C can be dropped onto the preparation immediately after removal of the coverslip, before drying. This may be helpful in dispersing the cytoplasm.

Acknowledgments

This work was supported by Sonata Bis 2020/38/E/NZ9/00033 project of the Polish National Science Centre and International Atomic Energy Agency, IAEA, Austria, under the framework of RER5024.

References

1. Maravilla AJ, Rosato M, Rosselló JA (2023) Preparation of mitotic chromosomes with the squash technique. In: Heitkam T, Garcia S (eds) Plant cytogenetics and cytogenomics. Methods in molecular biology, vol 2672. Humana, New York
2. Sugiyama M, Norizuki M, Kikuchi S, Yasui Y, Matsui K (2023) Development and

chromosomal characterization of interspecific hybrids between common buckwheat (*Fagopyrum esculentum*) and a related perennial species (*F. Cymosum*). Breed Sci 73:230–236

3. Östergen G, Heneen WK (1962) A squash technique for chromosome morphological studies. Hereditas 48:332–329
4. Schwarzacher T, Heslop-Harrison P (2000) Practical in situ hybridization. BIOS Scientific Publishers, Oxford
5. Quisenberry KS (1927) Chromosome numbers in buckwheat species. Botanical Gazette 83(1): 85–88. The University of Chicago Press

Chapter 9

Embryo Rescue Technique in Buckwheat *Fagopyrum esculentum* Moench

Galina Suvorova

Abstract

Culture of immature buckwheat embryos is an excellent explant for tissue culture and regeneration in vitro, as well as a method of embryo rescue in case of fruit abortion especially in interspecific crosses. Here, we describe the method of immature embryo rescue using a differentiated approach to different age groups. The method involves the development of embryos along the path of direct embryogenesis bypassing the callus stage.

Key words Embryo rescue, Buckwheat, Ovule, Ovary, Embryogenesis, Plantlet, Seedling

1 Introduction

Immature embryos as explants with high regeneration capacity are very convenient objects for culture in vitro. The first who successfully grew buckwheat embryos not younger than 15 days old on White medium was V. Poddubnaya-Arnoldi in 1948 [1]. Processes of differentiation and dedifferentiation in the culture of immature embryos on the media of a more rich composition were studied in detail later [2, 3]. The embryo rescue was successfully used to overcome interspecific incompatibility in crosses of common buckwheat *Fagopyrum esculentum* with other species of *Fagopyrum*. Interspecific hybrids were obtained in combinations *F. esculentum* x *F. cymosum* [4–6], *F. esculentum* x *F. homotropicum* [7], and *F. tataricum* x *F. esculentum* [8, 9]. Though the hybrid plants were obtained, most were sterile, and incompatibility was partially overcome. Buckwheat as a crop is characterized by many problems that have yet to be solved: susceptibility to biotic stresses such as ground frost, low water supply, drought, and photoperiod, which cause flowering and embryo abortions [10]. So embryo rescue technique may be used in interspecific and intraspecific crosses.

Alexander Betekhtin and Artur Pinski (eds.), *Buckwheat: Methods and Protocols*, Methods in Molecular Biology, vol. 2791,
https://doi.org/10.1007/978-1-0716-3794-4_9,

We have developed a method of buckwheat embryo rescue, a distinctive feature of which is a differentiated approach to different age groups of embryos. We minimized intermediate stages and excluded the callus tissue formation that allowed obtaining plants from immature embryos, quickly preserving the initial embryo genotype.

2 Materials

Prepare stock solutions of minerals and growth regulators for nutrient media using distilled water. Store all the concentrates at 4 °C.

2.1 Nutrient Media Composition

1. Mineral composition of MS medium: 1650 mg/L of NH_4NO_3, 1900 mg/L of KNO_3, 370 mg/L of $MgSO_4$ $7H_2O$, 170 mg/L of KH_2PO_4, 440 mg/L of $CaCl_2{\cdot}2H_2O$, 27.8 mg/L of $FeSO_4$ $7H_2O$, 37.3 mg/L of $Na_2EDTA{\cdot}2H_2O$, 0.025 mg/L of $CoCl_2$ $6H_2O$, 0.025 mg/L of $CuSO_4$ $5H_2O$, 8.6 mg/L of $ZnSO_4$ $7H_2O$, 0.25 mg/L of Na_2MoO_4 $2H_2O$, 0.83 mg/L of KI, 6.2 mg/L of H_3BO_3, 22.3 mg/L of $MnSO_4$ $4H_2O$ [11].
2. Mineral composition of B5 medium: 2500 mg/L of KNO_3, 250 mg/L of $MgSO_4$ $7H_2O$, 150 mg/L of NaH_2 PO_4 H_2O, 134 mg/L $(NH_4)_2$ SO_4, 150 mg/L of $CaCl_2{\cdot}2H_2O$, 27.8 mg/L of $FeSO_4$ $7H_2O$, 37.3 mg/L of $Na_2EDTA{\cdot}2H_2O$, 0.025 mg/L of $CoCl_2$ $6H_2O$, 0.025 mg/ L of $CuSO_4$ $5H_2O$, 2 mg/L of $ZnSO_4$ $7H_2O$, 0.25 mg/L of Na_2MoO_4 $2H_2O$, 0.75 mg/L of KI, 3 mg/L of H_3BO_3, 10 mg/L of $MnSO_4$ $4H_2O$ [12] (*see* **Note 1**).
3. Recommended concentrations of mineral stock solutions: macro salts 1:40, micro salts 1:1000, $FeSO_4$ $7H_2O$ and $Na_2EDTA{\cdot}2H_2O$ together (dissolve $Na_2EDTA{\cdot}2H_2O$ in hot water, and add $FeSO_4$ $7H_2O$) 1:200, $CaCl_2{\cdot}2H_2O$ 1:1000, vitamins together 1:1000, glycine 1:250.
4. Organic composition of all media: 10 mg/L of thiamin HCL, 1 mg/L of pyridoxine HCL, 1 mg/L of nicotinic acid, 2 mg/L of glycine, 30 g/L of sucrose, 100 mg/L of inositol, 1000 mg/L of casein hydrolyzate, 6 g/L of agar.
5. Growth regulator composition: induction media 1–2.25 mg/L of BAP (6-benzylaminopurine) + 0.2 mg/L of IAA (indole-3-acetic acid), induction media 2–2.25 mg/L of BAP + 2 mg/L of IAA, induction media 3–2 mg/L of IAA, regeneration media 0.6 mg/L of BAP, rooting media 1 mg/L of IBA (indole-3-butyric acid).

6. Recommended concentration of stock solutions of growth regulators: BAP 0.2 mg/mL, auxins 0.5 mg/mL. Dissolve BAP in 1–2 mL of HCl, and dissolve auxins in 1–2 mL of NaOH previously.

2.2 Sterilizing Solution

Prepare 0.25% water solution of sodium chlorhexidine digluconate immediately before use.

2.3 Instruments and Equipment

Stereo microscope, forceps, dissecting needles, glass tubes for tissue culture, laminar box, autoclave, analytical balances, distillatory, pH meter, automatic pipets, hotplate, refrigerator.

3 Methods

3.1 Preparing of Nutrient Media

1. Prepare nutrient media at room temperature. Take a glass and pour water less than you need. Add all components sequentially: macro and micro salt solutions, Fe and Ca solutions, vitamin solution, glycine solution, and growth regulator solutions. Weigh the appropriate amount of sucrose, inositol, casein hydrolysate, and agar, and add all components except agar to the medium. Add water to the final volume and dissolve all the components.
2. Adjust pH to 5.8–6.0 by dropping NaOH or HCl.
3. Place the glass with media on the hotplate, add the agar, and heat the medium to a boil, mixing all the time.
4. Pour hot medium into culture tubes and cover with foil lids.
5. Autoclave the medium at 1.2 kgf/cm for 20 min. The sterile medium can be stored at room temperature for several days.

3.2 Culture of 4–6 Days' Embryos (I Group)

All manipulations with buckwheat embryos and tissues are carried out in sterile air. Culture of buckwheat tissues are maintained at 25 °C under fluorescent light with the light intensity of 2000–3000 lux, day length of 16 h.

1. Take ovaries from plants immediately before inoculation on the medium. Clear them from the remnants of perianth. Sterilize in 0.25% sodium chlorhexidine digluconate solution for 5 min, and rinse three times in sterile water (*see* **Note 2**).
2. Excise the ovule from the ovary under stereomicroscope using a dissecting needle and forceps (*see* **Note 3**). Using the tip of the needle, put the ovule on the surface of the induction medium 1 in culture tube. Inoculate every ovule in a separate tube. The cover of the ovule dies and breaks, and the embryo begins to develop (*see* **Note 4**). At the young age, the embryos usually develop abnormally, forming abnormal seedling (Fig. 1) or morphogenic tissues (Fig. 2).

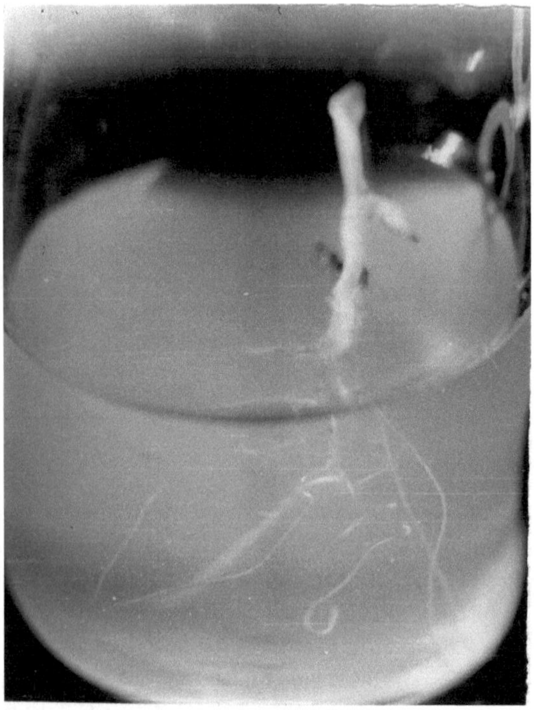

Fig. 1 Abnormal seedling formed from 7-day embryo

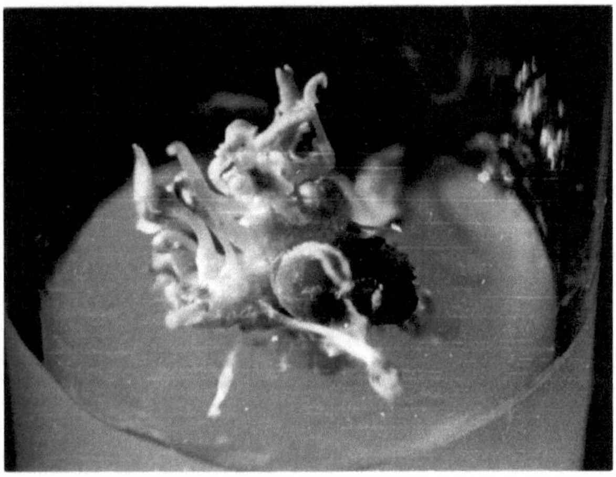

Fig. 2 Morphogenic tissue formed from 6-day embryo

3. Transfer the developed tissue to the regeneration medium after a month. Pass the growing tissues on regeneration medium every month. Separate the morphogenic tissue with emerging shoots from maternal tissue, and transfer to the hormone-free medium (*see* **Note 5**).
4. Cut off formed shoots with length of 3 cm from morphogenic tissue, and put them on the rooting medium. After 2 weeks, pull out the plantlets with well-formed roots from the agar,

Fig. 3 Buckwheat plantlet with washed roots in water

wash under running water, and place in a glass with small amount of water for 1–2 days (Fig. 3). Do not use distilled water in this case.

3.3 Culture of 7–10 Days' Embryos (II Group)

1. Isolate and sterilize the ovaries as described in points 1 and 2 of the previous section.
2. Put the isolated ovules separately on the induction medium 2 (*see* **Note 6**). In the case of 7–10-day-old ovules the embryos develop into seedling more often, but seedling is small and weak. Transfer the weak seedling to regeneration media or hormone-free media after a month. Pull out the formed seedling from media and place in glass with small amount of water.
3. If seedling has a weak thin root, cut the upper part of seedling and transfer it to the rooting media. After 2 weeks, remove the plantlets with well-formed roots from the agar, wash under running water, and place in a glass with small amount of water.

3.4 Culture of 11–15 Days' Embryos (III Group)

1. Take and sterilize the ovaries as described in points 1 of Subheading 3.2.
2. Excise the ovules from the ovaries on the surface of a sterile Petri dish using a dissecting needle and forceps (*see* **Note 7**).

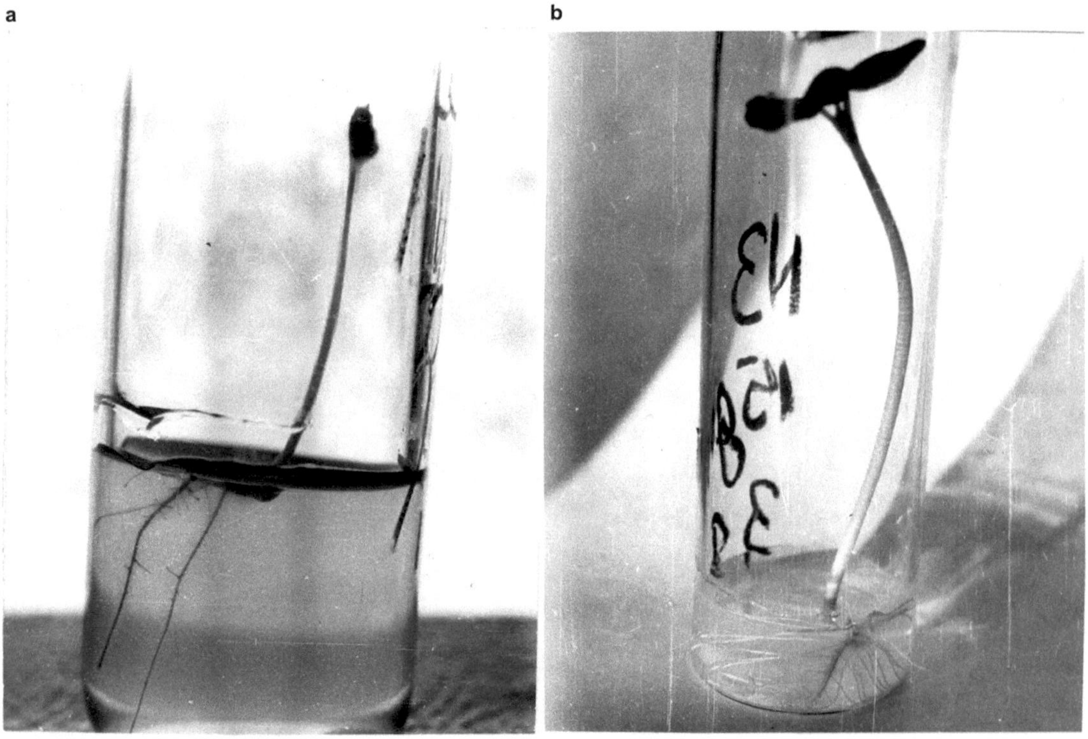

Fig. 4 (**a**) Eleven-day embryo and (**b**) 15-day embryo germination on induction medium 3

3. Sterilize ovules again in 0.25% sodium chlorhexidine digluconate solution for 5 min, and rinse three times in sterile water (*see* **Note 8**). Put the ovules each separately on the surface of the induction medium 3 using the forceps.
4. When seedlings are almost formed (Fig. 4a, b), pull them out from agar, and wash under running water, and place in a glass with a small amount of water (*see* **Note 9**).

3.5 Seedling Acclimatization

1. Take the plantlets or seedlings out of the water, and plant them in pots of sifted soil in the greenhouse. The recommended photoperiod is 18 h.
2. Cover the plants with a glass cup to create a humid chamber for the first few days, opening it occasionally. Remove the glass as soon as the plants grow (*see* **Note 10**).

4 Notes

1. Our experiment demonstrated that the development direction of embryos and their survival do not depend on the mineral composition of the nutrient medium. Both MS basal medium and B5 medium may be used with equal results.

2. For 4–6-day ovules and 7–10-day ones, single sterilization with a low concentration of sterilizing agent is recommended. It does not give 100% sterility, and there may be contamination up to 20–40%. But if we increase the sodium chlorhexidine digluconate solution concentration or increase the sterilization time, most of the explants will be necrotic. Young buckwheat ovaries are very tender. Under no circumstances should ethanol be used.
3. Buckwheat ovaries of 4–6 days' old are light green and trihedral, reaching 50–70% of the length of a mature fruit. Buckwheat ovules of 4–6 days old are translucent, club-shaped, and 2–3 mm long, and the diameter of the thickened part reaches 1 mm. Therefore, it is recommended to isolate them very carefully from the ovaries under a stereomicroscope so as not to damage them. In case of damage, callus tissue may form or the embryo may die. A group of 4–6-day ovules are very sensitive to the inoculation procedure in vitro. In our experiment, 10–20% of embryos of the first age group developed as seedlings, 60–70% developed abnormally, and 10–20% formed callus tissue.
4. The primary explant in our experiments is the ovule, since it is easier to isolate ovule without damaging a young embryo. But we use the term "embryo rescue," since during the first passage, the embryo germinates through the ovule coat dying off later.
5. In order to receive the initial embryo genotype without any somaclonal variations, we developed a regeneration protocol from morphogenic tissue but not from callus.
6. On the tenth day of buckwheat fruit growth, the ovary reaches the length of mature seed, but the embryos grow more slowly than the pericarp, and the 7–10-day-old ovaries look unfulfilled. Ovules of 7–10 days' old have an elongated trihedral shape of 3.5–4.0 mm long. If the procedure of sterilization of 7–10 days' ovules is similar to the previous, medium recommended is different. Medium 2 contains a large proportion of auxin. The embryos of the second age group have a higher survival rate than the youngest embryos, but more suffer from an infection.
7. The period of fruit formation in buckwheat *F. esculentum* averages 22 days. Mature buckwheat fruit is a trihedral nut in which in the center of it is the embryo with two curved cotyledons located surrounded by tightly fitting endosperm. Buckwheat ovaries of 11–15 days' old reach the size of a mature fruit but still have a green color. The 11-day ovaries are not completely fulfilled, and the 15-day ones contain an already formed embryo with a semiliquid endosperm. The ovules of the third growth group are formed enough to be carefully isolated without a microscope.

8. As the fruit develops, the infection penetrates the pericarp. Therefore, we recommend double sterilization: the first is ovary, while the second is the ovule. Double sterilization efficiency is 90–100%.
9. The embryos of 10–15 days' old develop through direct embryogenesis on a medium 3 which contains auxin only. After 1–2 weeks, most embryos form seedlings with a developed root system, light green hypocotyl, and large green cotyledons. Seedlings are washed from agar and planted into the soil.
10. The methods of embryo rescue are varied depending on embryo age. Plantlet regeneration from morphogenic tissue is recommended for young embryos, and direct embryogenesis occurs in the group of more mature embryos. If the embryo is older, then survival in vitro is higher. Therefore, we recommend rescuing more mature embryos in case of intraspecific crosses and taking very young embryos in case of incompatible interspecific crosses when hybrid embryos degenerated within some days after pollination.

References

1. Poddubnaya-Arnoldi V (1948) Comparative – embryological studies of diploid and tetraploid forms of buckwheat. Botanical J 33(2): 181–194. (in Russian)
2. Neskovic M, Vujicic R, Budimir S (1987) Somatic embryogenesis and bud formation from immature embryos of buckwheat (*Fagopyrum esculentum* Moench.). Plant Cell Rep 6: 423–426
3. Rumyantseva N, Salnikov V, Fedoseeva N, Lozovaya V (1992) Peculiarities of morphogenesis in long-term cultivated buckwheat calluses. Russ J Plant Physiol 39:143–151
4. Ujihara A, Nakamura Y, Minami M (1990) Interspecific hybridization in genus *Fagopyrum* – properties of hybrids (*F. esculentum* Moench. x *F. cymosum* Meissner) through ovule culture. Gamma Field Symposia Japan 29:45–53
5. Suvorova GN, Fesenko NN, Kostrubin MM (1994) Obtaining of interspecific buckwheat hybrid (*Fagopyrum esculentum* Moench. x *F. cymosum* Meissn.). Fagopyrum 14:13–16
6. Rumyantseva N, Fedoseeva N, Abdrakhmanova G, Nikolskaya V, Lopato S (1995) Interspecific hybridization in the genus *Fagopyrum* using *in vitro* embryo culture. In: Proceeding of the 6th International Symposium on buckwheat Japan, pp 211–220
7. Campbell C (1995) Inter-specific hybridization in the genus *Fagopyrum*. In: Proceeding of the 6th International Symposium on buckwheat Japan, pp 255–263
8. Samimy C, Bjorkman T, Siritunga D, Blanchard L (1996) Overcoming the barrier to interspecific hybridization of *Fagopyrum esculentum* with wild *Fagopyrum tataricum*. Euphytica 91:323–330
9. Wang Y, Scarth R, Campbell C (2002) Interspecific hybridization between *Fagopyrum tataricum* (L.) Gartn. and *F. esculentum* Moench. Fagopyrum 19:31–35
10. Tomasiak A, Zhou M, Betekhtin A (2022) Buckwheat in tissue culture research: current status and future perspectives. Int J Mol Sci 23: 2298. https://doi.org/10.3390/ijms23042298
11. Murashige T, Skoog F (1962) A revised medium for rapid growth and bioassays with tobacco tissue cultures. Physiol Plant 15(13): 473–497
12. Gamborg OL, Miller RA, Ojima K (1968) Nutrient requirement of suspension cultures of soybean root cells. Exp Cell Res 50(1): 151–158

Chapter 10

Comprehensive Embryological Analyses of Common Buckwheat (*Fagopyrym esculentum* Moench)

Aneta Słomka, Klaudia Sychta, and Agnieszka Płażek

Abstract

Knowledge of detailed reproductive biology of cultivated species is important as requirements for fruit and seed production allow the development of effective management strategies and a sustainable use. Embryological processes of common buckwheat (*Fagopyrum esculentum* Moench) are difficult to interpret due to the influence of genetic determinants, i.e., dimorphic heterostyly resulting in the production of long- and short-styled flowers, and environmental predisposition, i.e., sensitivity of ovules to thermal stress. Furthermore, the situation is complicated by overproduction of flowers and depletion of resources as the plant ages. Herein we provide protocols that allow to visualize both basic and more specific embryological features and also disturbances in sexual reproduction of common buckwheat resulting from external and internal factors. All stages of plant material fixation, preparation, staining, and observation are described and explained in detail. Technical tips and pictures of properly prepared microscopic sections are also provided.

Key words Female gametophyte, Pollen tubes, Embryo, Paraffin method, Pollen stainability by acetocarmine and Alexander dye, Aniline blue staining for callose

1 Introduction

Knowledge of detailed reproductive biology of cultivated species is important as fruit and seed production requirements allow the development of effective management strategies and sustainable use. Common buckwheat (*Fagopyrum esculentum* Moench) reproductive biology is complicated mostly due to the self-incompatibility which is an obstacle to establishing pure lines and fixation of agronomically useful genes. Reciprocal positioning of stigmas and anthers in two different floral morphs (heterostyly) promotes compatible pollination in this species [1]. Pollen of short-styled flowers (Thrum; S/s) fertilizes long-styled flowers (Pin; s/s) and vice versa (Fig. 1a). Illegitimate germination is also possible; however, pollen tubes do not reach the embryo sac; thus, fertilization fails. Legitimate pollen tubes reach as far as beyond two-thirds of the style length which could be easily visualized by

Alexander Betekhtin and Artur Pinski (eds.), *Buckwheat: Methods and Protocols*, Methods in Molecular Biology, vol. 2791,
https://doi.org/10.1007/978-1-0716-3794-4_10,

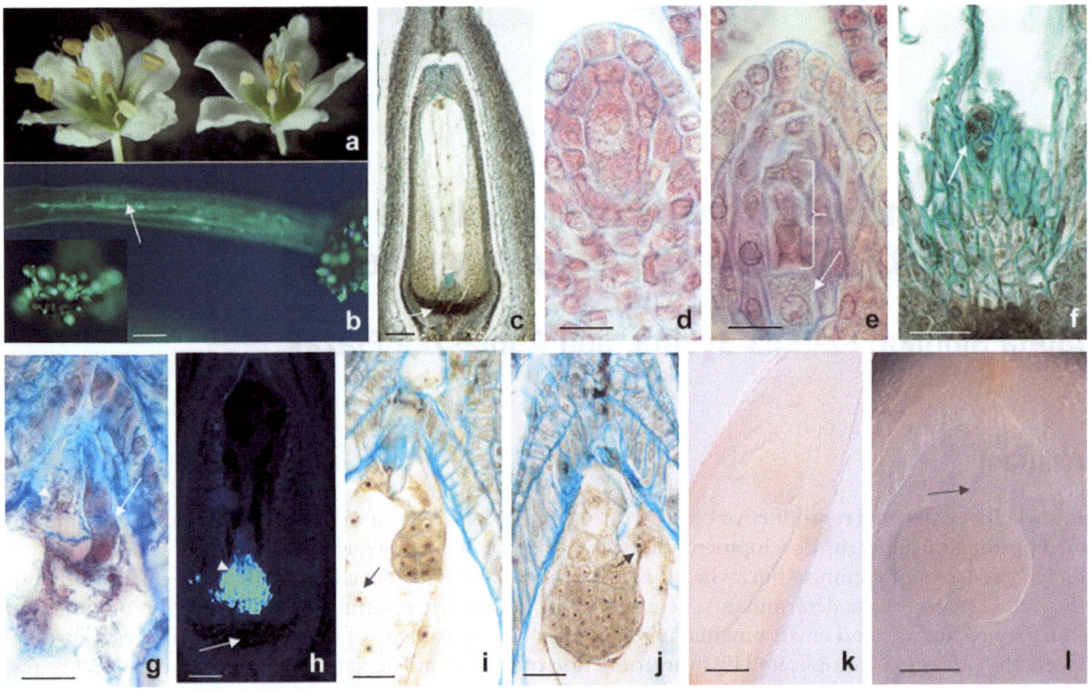

Fig. 1 Embryological features of common buckwheat – different developmental stages revealed by various preparation techniques and staining methods. Two flower morphs, Thrum (left) and Pin (right) varying in the length of the style and stamens (**a**). Pollen grains (inserted) germinating on the stigma of Pin flower; yellow-green fluorescent pollen tubes penetrating the style marked with arrow (**b**). Orthotropous, bitegmic, crassinucellate ovule; hypostase of the ovule marked with arrow (**c**). Megaspore mother cell at the prophase I (**d**). Linear tetrad of megaspores; three marked with bracket at the micropyle pole, i.e., degenerating, one marked with arrow at the chalazal pole, i.e., functioning (**e**). Podium of the ovule with three antipodal cells marked with arrow (**f**). Double fertilization; pollen tube penetrating one of two synergids marked with arrow (**g**). Podium of the ovule with cell walls covered with callose marked with arrowhead and hypostase of the ovule marked with arrow (**h**). Nuclear endosperm marked with arrows at the stages of a few-celled (**i**) and multicelled (**j**) globular embryo. Globular embryos (**k**) and the initial stage of the heart-shaped embryo (**l**); suspensor marked with arrow. Staining with aniline blue for callose (**b**, **h**), paraffin slides (**c**–**j**) stained with Heidenhain's iron hematoxylin combined with alcian blue (**c**, **f**, **i**, **j**) and Ehrlich's hematoxylin combined with alcian blue (**d**, **e**, **g**), ovules cleared with methyl salicylate (**k**, **l**). Bars in **a**, **f**, **h**–**j** = 25 μm; in **b**, **c**, **k**, **l** = 100 μm; in **d**, **e**, **g** = 10 μm (Picture **c** taken from [2], modified)

aniline blue staining (Fig. 1b). In some Polish cultivars and strains, severely biased ratio (1:2.7–1:19) of Pin and Thrum flowers could occur; however, it does not affect seed set or male success in either floral morph [2]. Another impediment in the sexual reproduction of common buckwheat is overproduction of flowers, leading to resource exhaustion and flower dropping and seed (embryo) abortion (large vacuolized or collapsed cells of embryo are indicators of this process) due to starvation. (Cyto)embryological studies, though developed in the first half of the previous century, allow also now, in the era of genomic and transcriptomic studies, for monitoring changes that take place in the target experimental systems [3]. Experiments with removing lateral ramifications of common buckwheat showed that plants with a lower number of flowers (likely also self-finishing genotypes, if they existed) could achieve a higher frequency of properly developing embryo sacs and also set more seeds per plants [4].

General embryology and embryogenesis of *F. esculentum* have been known since years [e.g., 5, 6], and recently knowledge on the subject has been expanded to include details of ovule development [2, 4, 7, 8]. Ovules of common buckwheat are orthotropous, bitegmic, crassinucellate (Fig. 1c), tetrad of megaspores linear (Fig. 1d, e); embryo-sac development *Polygonum*-type; polar nuclei fusing prior to fertilization; antipodal cell ephemeral (Fig. 1f); synergids pear-shaped or hooked with filiform apparatus (Fig. 1g); hypostase and podium present (Fig. 1c, h); endosperm formation nuclear (Fig. 1i and j); and embryogenesis *Asterad*-type (Fig. 1i-l). However, recent embryological studies revealed that female success of both Pin and Thrum flowers measured as regularity in the run of the embryological processes is dramatically low; as many as over 50% of ovules (dependently on the strain/cultivar) exhibit signs of degeneration (ovules, whole buds) or abortion (mature embryo sacs, proembryos, embryos) [2, 7, 8]. Two factors, usually in combination, could affect the run of embryological processes in female line of *F. esculentum*: sensitivity to thermal stress and resource limitations as the plant ages. A drastic reduction in the number of properly developed embryo sacs was thus observed in open flowers cultivated at 30 °C compared to 20 °C [7]. Similarly, in vitro cultivated flowers on media with strongly reduced nutrient content demonstrated a drastic increase in degenerated embryo sac frequency. Furthermore, in *in planta* experiments, when 50% of flowers or all lateral ramifications were removed, the number of regularly developing ovules and setting seeds was significantly improved [8].

Embryological studies of ovules unlike pollen grains are much more time-consuming as embryo sacs (=megagametophytes, female gametophytes) are located within ovules and wrapped by integuments. In *F. esculentum* only a single ovule and thus a single female gametophyte develops within an ovary, which requires observation of many flowers (ovaries) to interpret pictures contrary to species with ovary developing dozens of ovules inside. Perhaps this is the reason why disturbances in female gametophyte development in common buckwheat have been confirmed relatively recently. Previously, only disturbances such as abortive or abnormal embryos had been indicated [see 2 and lit. cited therein]. Contrary to ovules, pollen grains production in *F. esculentum* is not influenced by environmental factors, e.g., high temperature. The overall quality of pollen in common buckwheat is high reaching above 90% of stainability by Alexander's and acetocarmine tests, rarely falling below this number [2].

Proceeding multiple embryological analyses at the same time is unfeasible; thus, it is recommended to fix plant material in order to use it at a convenient time. It allows starting the whole procedure even after many years since plant material collection. Both fixatives AA (aceto-alcohol) and FAA (formalin aceto-alcohol) described below are suitable for common buckwheat flowers, ensuring comparable results. Clearing of whole ovules (Fig. 1k and l, *see* **Note 1**), their sectioning (cryo-freeze or embedded in resins, paraffin, or waxes) allows the internal insight of the ovule. Herein, we present the paraffin method. Three main steps in the embryological sample (just ovule or ovule plus stamens) preparation of *F. esculentum* are described: (1) tissue embedding in paraffin including fixation, dehydration, clearing, infiltration, and embedding; (2) microtome sectioning and affixing slides to the microscopic slide; (3) staining (with or without differentiation step) of the sections for visualization under the light microscope. Staining of pollen tubes for callose, also described herein, requires: (1) softening and clearing pistil and ovary (in an elevated temperature to speed up tissue maceration); (2) rinsing and venting; and (3) staining. Pollen grain (=microgametophyte, male gametophyte) observation relies simply on: (1) pollen (fresh or fixed) isolation; and (2) staining.

We hope that the availability of this protocol for common buckwheat will allow for a considerable saving of time. It may be useful for those researchers with limited knowledge and experience in this field whose topic requires the application of embryological procedures to keep moving forward in their investigation. Keep in mind that observation and interpretation of the embryological sections usually requires expertise which could be facilitated by the use of specific books (atlases) [e.g., 9, 10], devoted embryology including embryogenesis.

2 Materials

2.1 Solutions and Stains

1. Mayer's albumin: glycerin and hen egg white (1:1; v:v), few crystals of thymol. Filter through glass wool and keep in 4 °C.
2. Ehrlich's hematoxylin: hematoxylin 2 g, absolute ethanol 100 mL, glycerol 100 mL, distilled water 100 mL, glacial acetic acid 10 mL, aluminum potassium sulfate ($AlK(SO_4)_2 \cdot 12H_2O$) – to excess.
3. Alcian blue stain: alcian blue 100 mg, glacial acetic acid 3 mL, distilled water 97 mL.
4. Heidenhain's iron hematoxylin: hematoxylin 0.5 g, 96% ethanol 10 mL, distilled water 90 mL (the solution should be maintained on light for at least 2 weeks prior using).

5. Ferric solution: ferric ammonium sulfate ($NH_4Fe(SO_4)_2 \cdot 12H_2O$) 4 g, distilled water 100 mL.
6. Alexander's stain solution according to [11]:
   ethanol 95% 10 mL, 1% malachite green in 95% ethanol 1 mL, distilled water 50 mL, glycerol 25 mL, phenol 5 g, chloral hydrate 5 g, 1% acid fuchsin in water 5 mL, 1% orange G in water 0.5 mL, glacial acetic acid 2 mL.

2.2 Equipment

1. Heat plate.
2. Microtome.
3. Incubator.

3 Methods

3.1 Paraffin Method for Observation of Microsporogenesis and Megasporogenesis Run, Pollen and Female Gametophyte Development, and Embryo and Endosperm Formation

3.1.1 Fixation and Sections Preparation

1. Fix flowers at different developmental stages (e.g., buds, fully developed, wilted flowers) (*see* **Note 2**) immediately after the collection in a freshly prepared AA fixative (glacial acetic acid and 96% ethanol, 1:3; v:v) and/or FAA solution (formalin - formaldehyde aqueous solution, glacial acetic acid, 96% ethanol, deionized water, 2:1:10:7; v:v:v:v) for at least 24 h. It is important to use pure ethanol and fresh formaldehyde; the latter should be kept in dark. Be careful as formaldehyde is human carcinogenic. The fixatives should be prepared directly before use, and their volume should be at least two to three times higher than the volume of the fixed flowers. At this stage, petals could also be removed, optionally. Flowers should also be slowly shaken during fixation or be fixed in a vacuum desiccator.
2. Dehydrate fixed flowers in a graded ethanol series (70%, 96%, 100% for 24 h of each), and saturate them with grading concentration of chloroform (chloroform/absolute ethanol; 1:3, 1:1, 3:1, 1:0; v:v, each for 2–3 h at room temperature).
3. Leave a bit of chloroform, slightly above the level of flowers, add paraffin beads, and place unsealed bottles with flowers, chloroform, and paraffin into the incubator in a temperature which does not exceed the temperature of the paraffin melting more than 2–3 °C. For common buckwheat, selected for analyses paraffin does not need to be hard (of high-temperature melting point); melting point between 52 and 57 °C is accurate. After around 10 days, when all chloroform evaporates, flowers could be embedded in paraffin blocks. You can also keep them in liquid paraffin for a longer period; however, do not raise the temperature.
4. Using previously heated (to avoid rapid paraffin hardening) metal frames (serving as the edges) and glass plates (serving as

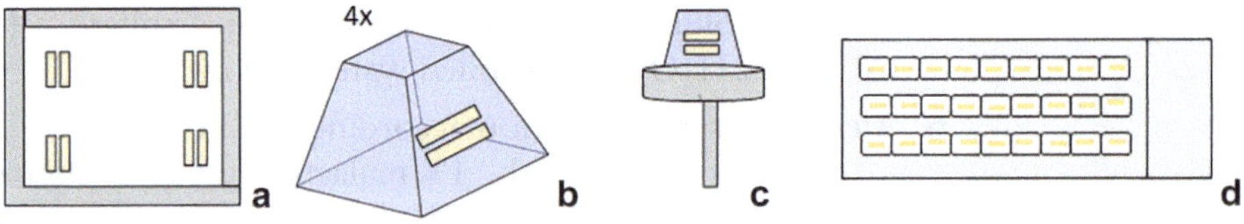

Fig. 2 Preparation of paraffin sections of common buckwheat flowers. The way of arranging flowers in a paraffin block - Two flowers arranged in four corners (**a**); cut out and trimmed four (from each corner) truncated pyramid (**b**); one of them attached with paraffin to the microtome holder (**c**); series of sections (ribbons) adhered with Mayer's albumin to the microscopic slide (**d**)

a bottom) or aluminum gutters (made of food-grade aluminum foil), prepare paraffin blocks. Put the flowers in parallel with a heated needle or tweezer, and fill the tank with liquid paraffin (Fig. 2a). No more than two to three flowers, so that the length of the paraffin block after cutting out (taking into account the excess paraffin needed for fastening to the holder) does not exceed 1.5 cm, should be put next to each other (Fig. 2b).

5. When paraffin blocks are solid, remove the frames and the excess of the paraffin – trim each block and fasten it to the microtome holder. A single block is a truncated square or rectangular pyramid. The base, which will be attached to the holder, may be larger than the opposite one from which you will start cutting (Fig. 2b, c). Cut flowers longitudinally into ~15 μm sections on a rotary microtome. Tip: ensure the block is cold before cutting. Place the blocks on a cold plate or a cold wet surface for a few minutes before cutting.
6. Put serial cuts (ribbons) on microscope slide covered with a droplet of spread Mayer's albumin covered with the deionized water on top of it (both covering the whole surface of the microscopic slide; don't overdo it with the amount of the Mayer's albumin) (Fig. 2d). The microscope slide should be perfectly cleaned in 70% ethanol before putting Mayer's albumin and water on it.
7. Place the microscope slide onto the heating plate warmed up to 40 °C to expand ribbons thermally. Control the ribbon position with a dissecting needle and left microscopic slide on the heating plate until the whole water evaporates and the ribbons stretch.
8. Dry sections at 30 °C overnight, for example, on the roof of a incubator.

3.1.2 Double Staining of Paraffin Sections with Ehrlich's Hematoxylin and Alcian Blue

Prepare ahead a set containing 15 pieces of Coplin jars filled with xylene (4 pieces), ethanol at different concentrations (8 pieces), alcian blue stain (1 piece), Ehrlich's hematoxylin stain (1 piece), and distilled water (1 piece).

1. To remove paraffin, put microscope slides into two xylene changes (each for 10 min).
2. Hydrate slides with degraded ethanol series: 100%, 96%, 70%, 50% (for 10 min, 10 min, 8 min, and 5 min, respectively).
3. Rinse microscope slides with distilled water for 5 min, and then put them into alcian blue stain for 3 min.
4. Rinse slides with 50% ethanol and stain in Ehrlich's hematoxylin stain for 3–10 min (the observation of the staining intensity is needed as it may be changing dependently on, e.g., sections thickness), and next wash them in 50% ethanol again.
5. Finally, dehydrate slides with graded ethanol series – 70%, 96%, and 100% (for 8 min, 10 min, and 10 min, respectively) – and wash them twice with xylene for 10 min each.
6. Cover slides with a drop of Entellan™ mounting medium, and dry them for 24 h at room temperature prior to microscopic observation.

Ehrlich's hematoxylin stains nuclei purple or reddish purple, and alcian blue stains cell walls bright blue.

3.1.3 Double Staining of Paraffin Sections with Heidenhain's Iron Hematoxylin and Alcian Blue

Prepare ahead a set containing 17 pieces of Coplin jars filled with xylene (4 pieces), ethanol at different concentration (8 pieces), alcian blue stain (1 piece), Heidenhain's iron hematoxylin stain (1 piece), distilled water (1 piece), tap water (1 piece), 4% ferric solution (1 piece), and Perti dish with 1% ferric solution for differentiation.

1. Points 1–2 are the same as in Subheading 3.1.2.
2. Put slides into 4% ferric solution for 30 min. Ferric solution activates the carboxylic groups of proteins.
3. Rinse slides under running tap water for 5 min, and put into distilled water for a while.
4. Put slides into Heidenhain's iron hematoxylin and stain for 30 min. Hematoxylin is binding to the iron used in point 2.
5. Make slides differentiation (the differentiation solution removes the hematoxylin from the tissue until the staining intensity and quality are optimal):
   (a) Rinse slides in tap water and then with distilled water.
   (b) Put slides onto Petri dish and pour 1% ferric solution. The differentiation process should be controlled under the light microscope by adding drops of ferric solution to the tissue sections (*see* **Note 3**). It usually takes 1–5 min.
   (c) Rinse slides under running tap water intensely for 20 min, and rinse briefly in distilled water.

6. Move to the point 3 of Subheading 3.1.2, then rinse slides with 50% ethanol, and next move to points 5 and 6 of Subheading 3.1.2.

Heidenhain's iron hematoxylin stains nuclei dark blue, and alcian blue stains cell walls bright blue.

3.2 Alexander's and Acetocarmine Tests for Pollen Viability Analysis

1. Fix large flower buds in a mixture of ethanol and AA and/or FAA solution according to the procedure described in Subheading 3.1.1. Optionally unfixed pollen grains could also be stained.
2. Prepare with dissecting needles pollen grains from all stamens of the flower bud to the drop of Alexander's dye/acetocarmine solution (1% in water) placed ahead on a microscope slide.
3. Stir isolated pollen grains gently with the needle. Remove stamens' debris, and cover with coverslip to avoid air bubbles closing. Optionally add a drop of glycerin, and/or cover the edges of the coverslip with clear nail polish. Both actions prevent the preparation from drying out quickly.
4. Observe pollen grains under the bright field of light microscope. Preparations are suitable for observation for about 2 weeks if kept in 4 °C. Viable pollen grains in Alexander's dye are purple-red (the pollen protoplasts stain red by acid fuchsin) with green sporodermis, and nonviable pollen grains are green (the cellulose of pollen walls stains green by malachite green) without viable protoplasts. Viable pollen grains in acetocarmine dye are reddish (acetocarmine stains nucleic acids), while nonviable remain unstained.
5. Count the frequency of viable pollen for as many pollen grains as possible, at least 100 per 1 flower bud.

3.3 Staining for Callose in Pollen Tubes of Germinating Pollen Grains (According to [11])

1. Dissected pistils from previously fixed (see Subheading 3.1.1) open flowers after spontaneous or hand-pollination put into 1 M NaOH for 30 min at 60 °C for clearing and softening.
2. Rinse pistils three times with distilled water and stain with 0.1% aniline blue (dissolved in 0.1 M K_3PO_4) for 1 h at 60 °C.
3. Put pistils on microscopic glass in a drop of glycerin, coverslip.

Pollen tubes with callose (β 1,3-glucan), the major component of the pollen tube cell wall, emit bright blue fluorescence under fluorescence microscope equipped with UV-2A filter. Compatible pollen tubes are those whose length is longer than two-thirds of the style.

4 Notes

1. Clearing of ovules with methyl salicylate is suitable only for embryos. As refers to the ovules that are crassinucellate, the observation of uncut ovules is difficult due to their thickness. This method is thus not recommended for common buckwheat.
2. If you are beginning your adventure with embryological studies of common buckwheat, start with choosing both flower types to see the differences in the ovule length and size.
3. The staining method with tissue differentiation is better if you want to observe cell's details (cyto-embryology).

Acknowledgments

This work was financially supported by the Polish Ministry of Agriculture and Rural Development (project no. HORhn 078/PB/34/14) and the National Science Centre in Poland (project no. 2017/25/B/NZ9/00148).

References

1. Wu LY, Chang FF, Liu SJ et al (2018) Heterostyly promotes compatible pollination in buckwheats: comparisons of intraflower, intraplant, and interplant pollen flow in distylous and homostylous *Fagopyrum*. Am J Bot 105(1): 108–116
2. Słomka A, Michno K, Dubert F et al (2017) Embryological background of low seed set in distylous common buckwheat (*Fagopyrum esculentum* Moench) with biased morph ratios, and biostimulant-induced improvement of it. Crop Pasture Sci 68:680–690
3. Moreno-Sanz P, D'Amato E, Nebish A et al (2020) An optimized histological proceeding to study the female gametophyte development in grapevine. Plant Methods 1:16–61
4. Hornyák M, Słomka A, Sychta K et al (2020) Reducing flower competition for assimilates by half results in higher yield of *Fagopyrum esculentum*. Int J Mol Sci 21:8953
5. Stevens NE (1912) The morphology of the seed of buckwheat. Bot Gaz 53(1):59–66
6. Lee MS (1986) Studies on heterostylism, fertility, and embryological characteristic in buckwheat, *Fagopyrum esculentum*. Korean J Crop Sci 31:129–142
7. Płażek A, Słomka A, Kopeć P et al (2019) Effects of high temperature on embryological development and hormone profile in flowers and leaves of common buckwheat (*Fagopyrum esculentum* Moench). Int J Mol Sci 20(7):1705
8. Płażek A, Hura K, Hura T et al (2020) Synthesis of HSP-70 and HSP-90 in flowers of common buckwheat (*Fagopyrum esculentum* Moench) under thermal stress. Crop Pasture Sci 71(8):760–767
9. Johri BM, Ambegaokar KB, Srivastava PS (eds) (1992) Comparative embryology of angiosperms, vol 1-2. Springer, Berlin/Heidelberg
10. Batygina TB (ed) (2002) Embryology of flowering plants: terminology and concepts, vol 1-3. Science Publisher Inc., Enfield Playmouth
11. Alexander MP (1969) Differential staining of aborted and nonaborted pollen. Stain Technol 44(3):117–122

Chapter 11

Phenol-Based Protein Extraction Method for Plant Proteomic Studies

Przemysław Kopeć, Monika Krzewska, and Agnieszka Płażek

Abstract

The protein extraction method based on the phenol solution and combined with protein precipitation with ammonium acetate in methanol and purification in the same solution, and additionally in acetone and ethanol, is recommended for proteomic studies of plant tissues. The obtained protein samples do not require additional nucleic acid digestion and removal of interfering contaminations. The presented protocol was used to analyze the proteome of common buckwheat flowers and leaves.

Key words Phenol extraction, Proteome, Protein purification, Protein solubilization, Protein precipitation

1 Introduction

Cell proteins are heterogeneous groups of compounds that have different solubility. This issue should be considered during sample preparation for proteome analysis. The sample should contain as many proteins as possible, purified from other compounds. Lipids, phenolic compounds, polysaccharides, pigments, and nucleic acids, present in plant tissues, make it challenging to extract proteins and their subsequent processing.

In 1986 Hurkman and Tanaka [1] proposed a phenol-based method for protein extraction and precipitation with ammonium acetate in methanol. This method was recommended for qualitative and quantitative comparisons analysis of plant membrane protein patterns by two-dimensional gel electrophoresis (2DE). The advantages of using phenol extraction are minimization of protein degradation due to proteolysis during sample preparation, a high cleanup capacity, removing interfering compounds of proteomic analysis, and high protein yield. The disadvantage of this method is the time-consuming nature and harmfulness of the phenol solution. In protein extraction, buffer-saturated phenol with Tris is

Alexander Betekhtin and Artur Pinski (eds.), *Buckwheat: Methods and Protocols*, Methods in Molecular Biology, vol. 2791, https://doi.org/10.1007/978-1-0716-3794-4_11,

used to separate nucleic acids. This is a good method for simultaneous extraction of nucleic acids and proteins. For precipitation of extracted proteins, a combination of salt and water-miscible organic solvent is used. Methanol effectively precipitates most proteins, with the exception of proteins from acidic solutions. In this case, ammonium acetate solves this problem.

Here we demonstrated a phenol-based method of protein extraction for two-dimensional electrophoresis, which was established according to Hurkman and Tanaka [1] with modification of Hajduch et al. [2]. It was used to analyze the flower and leaf proteome of common buckwheat [3], the leaf proteome of barley [4, 5], and the anthers' protein profiles of triticale [6, 7]. The procedure consists of six steps: extraction, precipitation, purification, solubilization, quantification, and sample aliquoting.

2 Materials

2.1 Equipment

1. Centrifuge 5804 R with rotors S-4-72 and F-45-30-11 (Eppendorf, Hamburg, Germany).
2. Shaker VXR basic Vibrax (IKA, Staufen im Breisgau, Germany).
3. Vortex VX-200 (Labnet International, Edison, USA).
4. Fridge and freezer.
5. Pipettes: 100–1000 μL, 1–5 mL.

2.2 Solutions and Reagents

Prepare solutions using ultrapure water and analytical grade reagents, unless otherwise specified. Dispose of the solutions after use in accordance with applicable laws and good laboratory practices.

1. Phenol solution, Tris-HCl saturated (pH 7.9). Store in the dark at 4 °C to 6 months (*see* **Note 1**).
2. Extraction buffer: 100 mM Tris-HCl (pH 8.8), 0.9 M sucrose, 10 mM EDTA; store at 4 °C.

   Add just before extraction β-mercaptoethanol to achieve a final concentration of 0.4% (v/v) (*see* **Note 2**).
3. Precipitation solution: 0.1 M ammonium acetate in 100% methanol (*see* **Note 3**).
4. Acetone: 80% (v/v) solution in water (*see* **Note 3**).
5. Ethanol: 70% (v/v) solution in water (*see* **Note 3**).
6. Sample solution: 7 M urea, 2 M thiourea, 4% (w/v) CHAPS, 50 mM DTT, 0,002% (w/v) bromophenol blue. Store in 1.5 cm^3 aliquots at −20 °C.

3 Methods

Plant material, collect in 2.0-mL tubes with screw caps and silicone sealing suitable for low temperatures, immediately freeze in liquid nitrogen, lyophilize, and grind for powder.

Perform all steps at room temperature under a fume hood.

3.1 Protein Extraction

1. Mix 5 cm^3 cold extraction buffer and 5 cm^3 cold phenol solution in a 50-mL conical centrifuge tube with screw cap (*see* **Note 4**).
2. Transfer ~0.2 g of ground lyophilized tissue to tube with extraction buffer and phenol, vortex, and shake for 30 min at 4 °C.
3. Centrifuge the sample for 15 min at 5000 × *g* and 4 °C.
4. The phenolic phase, on the top of the tube, recovers carefully to avoid contact with the interphase, and pour into a new 50-mL conical centrifuge tube.

3.2 Protein Precipitation

1. Add 5 vol of ice-cold precipitation solution. Incubate at −20 ° C overnight (*see* **Note 5**).
2. Collect the precipitate by centrifugation for 10 min at 5000 × *g* and 4 °C.

3.3 Protein Purification

1. Wash the pellet with proteins twice with 6 cm^3 ice-cold precipitation solution, then twice with 6 cm^3 ice-cold 80% acetone, and finally once with 6 cm^3 ice-cold 70% ethanol. After each washing step, centrifuge the sample for 10 min at 2000 × *g* and 4 °C.
2. After the last centrifugation, discard the supernatant and transfer the pellet with a small amount of 70% ethanol to the 2.0-mL tube (*see* **Note 6**).

3.4 Protein Solubilization

1. Centrifuge the sample for 15 min at 5000 × *g* and 4 °C, and discard the supernatant.
2. Add 0.2 cm^3 sample solution, vortex, and shake vigorously for 1 h at 4 °C.
3. Centrifuge the sample for 20 min at 5000 × *g* and 4 °C, and transfer the supernatant with proteins to a new 2.0-mL tube.

3.5 Sample Aliquoting

1. Quantify the protein in the samples (*see* **Note 7**).
2. Prepare aliquots of sample so that the specified amount of protein is in them depending on further processing. Store at −50 °C or below.

4 Notes

1. Tris-saturated phenol can be made by yourself in the laboratory, or you can buy commercial products such as Phenol, saturated pH 6.6/7.9 (PHE510, Bioshop, Mainway, Canada). **Caution**. Phenol is toxic if swallowed, in contact with skin, or inhaled. It is suspected of causing genetic defects and may cause organ damage. Do all operations under a fume hood, wearing lab coat, nitrile gloves (the EN374 standard), and tightly fitting safety goggles (the EN166 standard). Phenol solution is useable when it is colorless. For extended storage of the Tris-saturated phenol, −20 °C with protection from light is recommended.
2. **Caution**. β-Mercaptoethanol is toxic if swallowed or inhaled. It causes skin irritation and serious eye damage. It may cause damage to organs. It is suspected of damaging the unborn child. Add it under a fume hood, wearing lab coat, nitrile gloves (the EN374 standard), and tightly fitting safety goggles (the EN166 standard).
3. Prechill solution at −20 °C overnight before use.
4. The phenol solution is capped with a layer of Tris buffer. Pipet the whole required volume at once to avoid mixed of phases. The end of the tips must pass buffer layer. If the phases were mixed, wait for 2–4 h before use for the phases to separate.
5. For protein extraction, 5 cm^3 of phenol solution is used, and the same amount of phenolic phase should be collected. In this case, add 20 cm^3 of precipitating solution.
6. Proteins suspended in 70% ethanol can be stored at −20 °C.
7. When protein samples are quantified, use a method that does not interfere with reagents used in the sample buffer (detergents, reductants, and chaotropes). Protein concentration can be determined using the 2-D Quant Kit (GE Healthcare, Chicago, USA), which is designed for protein quantitation in samples for electrophoresis.

References

1. Hurkman WJ, Tanaka CK (1986) Solubilization of plant membrane proteins for analysis by two-dimensional gel electrophoresis. Plant Physiol 81:802–806. https://doi.org/10.1104/pp.81.3.802
2. Hajduch M, Ganapathy A, Stein JW, Thelen JJ (2005) A systematic proteomic study of seed filling in soybean. Establishment of high-resolution two-dimensional reference maps, expression profiles, and an interactive proteome database. Plant Physiol 137:1397–1419. https://doi.org/10.1104/pp.104.056614
3. Kopeć P, Hornyák M, Pastuszak J, Szczerba A, Rapacz M, Waga J, Płażek A (2021) Changes in the flower and leaf proteome of common buckwheat (*Fagopyrum esculentum* Moench) under high temperature. Int J Mol Sci 22:2678. https://doi.org/10.3390/ijms22052678
4. Gołębiowska-Pikania G, Kopeć P, Surówka E, Janowiak F, Krzewska M, Dubas E, Nowicka A,

Kasprzyk J, Ostrowska A, Malaga S, Hura T, Żur I (2017) Changes in protein abundance and activity induced by drought during generative development on winter barley (*Hordeum vulgare* L.). J Proteome 169:73–86. https://doi.org/10.1016/j.jprot.2017.07.016

5. Gołębiowska-Pikania G, Kopeć P, Surówka E, Krzewska M, Dubas E, Nowicka A, Rapacz M, Wójcik-Jagła M, Malaga S, Żur I (2017) Changes in protein abundance and activity involved in freezing tolerance acquisition in winter barley (*Hordeum vulgare* L.). J Proteome 169:58–72. https://doi.org/10.1016/j.jprot.2017.08.019
6. Krzewska M, Gołębiowska-Pikania G, Dubas E, Gawin M, Żur I (2017) Identification of proteins related to microspore embryogenesis responsiveness in anther cultures of winter triticale (× *Triticosecale* Wittm.). Euphytica 213:192. https://doi.org/10.1007/s10681-017-1978-1
7. Krzewska M, Dubas E, Gołębiowska G, Nowicka A, Janas A, Zieliński K, Surówka E, Kopeć P, Mielczarek P, Żur I (2021) Comperative proteomic analysis provides new insights into regulation of microspore embryogenesis induction in winter triricale (× *Triticosecale* Wittm.) after 5-azacytidine treatment. Sci Rep 11:22215. https://doi.org/10.1038/s41598-021-01671-y

Chapter 12

Two-Dimensional Gel Electrophoresis in Studies of Flower and Leaf Proteome of Common Buckwheat

Przemysław Kopeć, Monika Krzewska, and Agnieszka Płażek

Abstract

Two-dimensional gel electrophoresis (2-DE) is a proteomic tool used for the separation of protein mixtures according to protein isoelectric point and molecular mass. Although gel-free quantitative and qualitative proteomic study techniques are now available, 2-DE remains a useful analytical tool. The presented protocol was performed to analyze the flower and leaf proteome of common buckwheat using 24 cm immobilized pH gradient strips (pH 4–7) and visualization of proteins on gels via colloidal Coomassie G-250 staining.

Key words Two-dimensional gel electrophoresis, Proteome analysis, Isoelectric focusing, SDS-PAGE, Colloidal Coomassie staining

1 Introduction

Two-dimensional gel electrophoresis (2-DE) is a method for the separation of protein mixtures extracted from cells, tissues, or other biological samples. It is the primary technique for proteomics work. The method is a combination of two orthogonal separation techniques: isoelectric focusing (IEF) and sodium dodecyl sulfate-polyacrylamide gel electrophoresis (SDS-PAGE). During the IEF, the first dimension, proteins are separated by their isoelectric point (pI). In SDS-PAGE, the second dimension, proteins are separated according to their molecular mass (Mr). 2-DE was first introduced by O'Farrell [1] and Klose [2] in 1975 and has undergone numerous modifications. Seven years later, IEF in immobilized pH gradients was introduced [3]. Although gel-free quantitative and qualitative proteomic study techniques are now available, 2-DE remains a useful analytical tool.

Here, we describe how the proteins were separated and visualized during the proteome analysis of common buckwheat flowers and leaves [4] (Fig. 1). We performed 2-DE using the equipment

Alexander Betekhtin and Artur Pinski (eds.), *Buckwheat: Methods and Protocols*, Methods in Molecular Biology, vol. 2791, https://doi.org/10.1007/978-1-0716-3794-4_12,

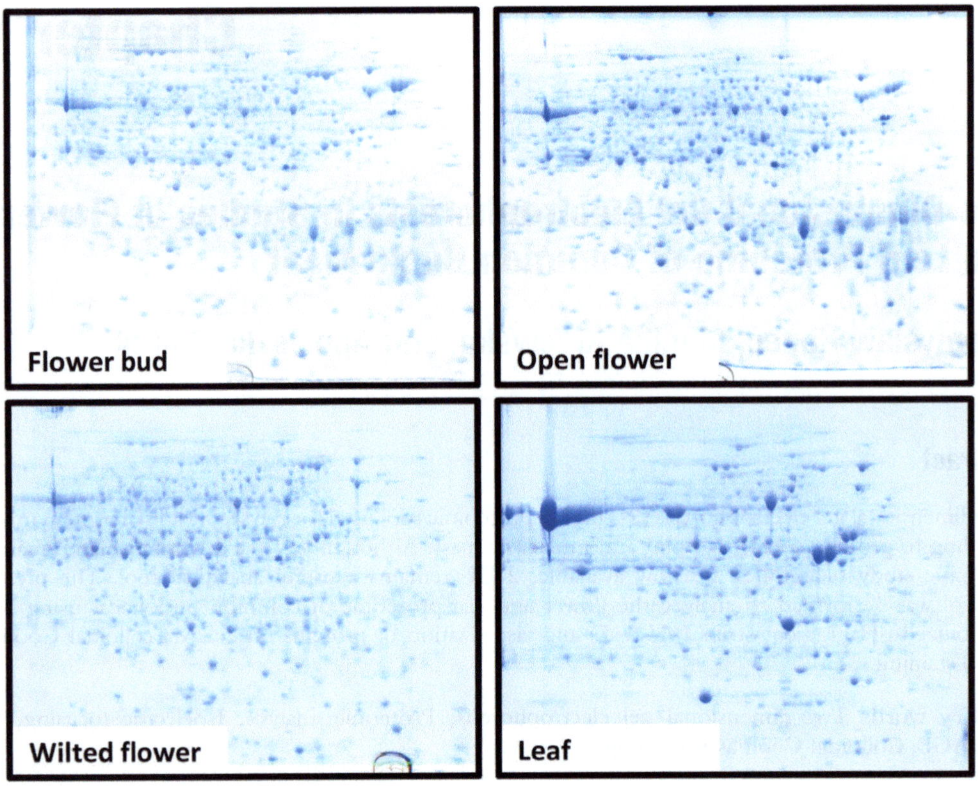

Fig. 1 Electropherograms present proteomes of flower bud, open flower, wilted flower, and leaf of common buckwheat (*Fagopyrum esculentum*) obtained by two-dimensional electrophoresis (2-DE) and colloidal Coomassie G-250 staining

that allows separation in the first dimension using 24 cm immobilized pH gradient (IPG) strips (pH 4–7), which are the longest available strips. After the first dimension, the IPG strips were equilibrated and transferred onto 12.5% SDS-PAGE gels. The use of medium-pH-range strips was preceded by trial protein separations using broad-pH-range strips (pH 3–11). The proteins on gels were visualized via colloidal Coomassie G-250 staining. The staining method has been modified from Neuhoff et al. [5]. Colloidal Coomassie G-250 staining, although less sensitive than other methods, is a relatively simple technique.

2 Materials

2.1 Equipment

1. Centrifuge 5804 R with rotors S-4-72 and F-45-30-11 (Eppendorf).
2. Ettan™ IPGphor™ 3 Isoelectric Focusing System (Cytiva).
3. Large format vertical gel electrophoresis unit SE900 (Hoefer).
4. Power supply PS600 (Hoefer).

5. Refrigerated circulator LTD6G (Grant) (*see* **Note 1**).
6. Heating block.
7. Rocking platform shaker 230 VAC (Labnet International).
8. Scanner Perfection V850 Pro (Epson).
9. IPGbox with Reswell Tray (Cytiva).
10. IPG strips: Immobiline® Drystrip pH 4–7, 24 cm (Cytiva).
11. Electrode wicks (Cytiva).

2.2 Solutions and Reagents

1. Sample buffer: 7 M urea, 2 M thiourea, 4% (w/v) CHAPS, 50 mM DTT, 0.002% (w/v) bromophenol blue; store in 1.5 cm^3 aliquots at −20 °C.
2. Carrier ampholytes, range pH 4–7 (e.g., IPG Buffer pH 4–7; Cytiva).
3. Mineral oil (e.g., Immobiline™ DryStrip Cover Fluid; Cytiva).
4. Gel solution: 12.5% acrylamide concentration, 1.5 M Tris-HCl (pH 8.8), 10% (w/v) SDS and water (*see* **Note 2**).
5. *N,N,N′,N′*-Tetramethylethylenediamine (TEMED).
6. Ammonium persulfate (APS): 10% (w/v) solution in water (*see* **Note 3**).
7. Isopropanol.
8. Gel storage solution: 1.5 M Tris-HCl (pH 8.8), 10% (w/v) SDS; store at 4 °C.
9. Equilibrium buffer: 1.5 M Tris-HCl (pH 8.8), 6 M urea, 30% (v/v) glycerol, 5% (w/v) SDS supplemented with DTT (80 mg per 4 cm^3) or with IAA (100 mg per 4 cm^3).
10. Agarose sealing solution: 25 mM Tris base, 192 mM glycine, 0.1% (w/v) SDS, 0.5% (w/v) agarose with some dissolved Coomassie Brilliant Blue (CBB) G-250 crystals; store in 2.0 mL aliquots at room temperature. CBB will be tracking dye.
11. Laemmli SDS electrophoresis buffer: 25 mM Tris base, 192 mM glycine, 0.1% (w/v) SDS; store at room temperature (*see* **Note 4**).
12. Colloidal Coomassie Blue G-250 working solution: 8% (w/v) ammonium sulfate, 0.8% (w/w) phosphoric acid, 0.08% (w/v) Coomassie Blue G-250, 20% (v/v) ethanol (*see* **Note 5**).

2.3 Software

1. Delta2D (Decodon).

3 Methods

3.1 IPG-Based IEF

1. Add 4.6 μL the carrier ampholytes to the appropriate amount of protein in the sample buffer, and fill up with the same buffer to 450 μL (*see* **Note 6**).
2. Transfer the solutions of proteins and carrier ampholytes in the sample buffer to the Reswell tray, and immediately put IPG strips in gel side down on the tray's channels. Place the Reswell tray in the IPGbox and gently close the lid. Allow the IPG strips to rehydrate overnight at room temperature (*see* **Note 7**).
3. Transfer IPG strips from the Reswell tray to the Manifold of Ettan IPGphor 3 IEF System; in gel side up, overlay the strips with mineral oil to minimize evaporation and prevent urea crystallization.
4. Wet the precut electrode wicks with deionized water, and place the pads on the ends of the IPG strips, position electrode assembly, and close Ettan IPGphor 3 IEF (*see* **Note 8**).
5. Start isoelectric focusing by applying a stepwise voltage gradient with a current limit of 75 μA per strip. Program for 24 cm IPG strips (pH 4–7): gradient 150 V for 30 min, hold 150 V for 2.5 h, hold 500 V for 1 h, gradient 10,000 V for 1 h, gradient 10,000 V for 3 h, and hold 10,000 for 55 min (*see* **Note 9**).
6. After the IEF is complete, proceed to the second-dimension separation immediately, or store IPG strips (*see* **Note 10**).

3.2 SDS-Page

1. Equilibrate IPG strips by gently shaking them twice for 20 min in equilibration buffer, first supplemented with DTT and then with IAA (*see* **Note 11**).
2. Remove excess equilibration buffer from the IPG strips by holding the end of the strip with forceps and lifting it and then placing it (gel side up) on a dust-free paper towel.
3. Remove residual liquid from the top of the slab gel by tipping the cassette and touching the liquid with a dust-free paper towel.

   Cast gels 1 day before use. Use instructions in the large format vertical gel electrophoresis unit SE900 to prepare the 12.5% SDS-PAGE gels. In short, add 2.4 cm^3 of 10% (w/v) APS and 126.8 μL TEMED to 400 cm^3 of gel solution. After gently mixing, pour the solutions to cassettes in Multiple Gel Caster. Overlay each gel with a thin layer of isopropanol. After 1 h, remove the gel cassettes from the caster, and store until the next day in gel storage solution at 4 °C.
4. Position of the equilibrated IPG strips on the cast gels (*see* **Note 12**). With a thin (<1 mm) plastic spatula, gently push the strip

down so that the entire lower edge of the strip is in contact with the top slab of the gel, without trapped air bubbles between them.

5. Overlay the agarose sealing solution on the IPG strips (*see* **Note 13**).
6. Insert the gel cassettes into the tank previously filled with the Laemmli SDS electrophoresis buffer, and replenish the buffer to the proper fill level. Turn on the refrigerated circulator pump with cooling to 25 °C and the circulating pump system in the tank.
7. Close the safety lid of the tank and connect the plugs to the power supply. Set the power supply for 80 mA/gel, and the voltage and wattage should be set to their maximums, 600 V and 100 W, respectively, so as not to limit the current (*see* **Note 14**).
8. Start the run. The end of the runs is when the tracking dye has reached the opposite side of the gels.

3.3 Gel Staining

1. Carefully remove each gel cassette. Do not remove the cassettes together with the PAGE rack from the tank. Place them on dust-free paper towels.
2. Use the plastic spatula to gently open the nonhinged side of the gel cassette. Spray the gel with deionized water, and carefully move the gel from the cassette into the box with deionized water.
3. Wash the gels three times in deionized water, rocking them for 15 min each time. The water in the box should cover the gels well.
4. After draining the water, after the last washing, pour into the box with gel the working solution Colloidal Coomassie Blue G-250. Stain the gels overnight, gently rocking them.

3.4 Gel Digitalization

1. Twice rinse gels with deionized water. Keep gels submerged in deionized water until image capture is complete.
2. Scan gels in 16-bit grayscale at a resolution of at least 300 dpi.
3. Process the gel images with Delta2D software.

4 Notes

1. In the recirculating bath, water is only used.
2. A ready-to-use acrylamide/bis-acrylamide solution is available (mixture of 30% (w/v) acrylamide/bis-acrylamide solution; 29:1), or it can be prepared by yourself. The gel solution can be deaerated prior to use using an oil vacuum pump. **Caution**.

The solution is harmful if swallowed or inhaled. It causes damage to the peripheral nervous system. It is suspected of damaging fertility and the unborn child. It is carcinogenic and mutagenic. Avoid the generation of vapors and aerosols. Do all operations under a fume hood, wearing lab coat, nitrile gloves (the EN374 standard), and tightly fitting safety goggles (the EN166 standard). It is good practice to wear two pairs of gloves when handling both the solution and the polyacrylamide gels.

3. Prepare the solution just before use. Fresh APS "crackles" during dissolution.
4. This solution can be prepared by diluting one volume of the tenfold concentrated solution with nine volumes of double-distilled water.
5. Mix Coomassie Blue G-250, ethanol, and 85% (w/v) phosphoric acid, and stir overnight. The next day, dissolve ammonium sulfate in water, add to Coomassie Blue G-250 solution, fill up with water to the required volume, and strain through the filter paper.
6. When using 24 cm IPG strips pH 4–7 and staining with Colloidal Coomassie Blue G-250 stain, the suitable sample load is from 200 to 1300 μg depending on the amount of extracted protein.
7. At the end of the sample transfer, perform a quick-run centrifugation of the tubes to avoid foaming and complete collection of the protein solution. The sample applied to the Reswell tray channel should be immediately covered with an IPG strip because the urea crystallization occurs quickly. During the rehydration of the IPG strip gel, proteins are also transferred to the gel.
8. Electrode wicks can be wet by immersing in water and draining the excess on a paper towel or applying 150 μL of water with a pipette.
9. The last step of IEF can be programmed: hold 150 V for any time. This prevents the proteins from diffusion off the gel until further processing.
10. IPG strips can be stored at −50 °C or below. The Reswell tray enclosed in the box can be used for storage.
11. IPG strips can be equilibrated on the Reswell tray, using 4 cm^3 of equilibration buffer per channel.
12. The molecular weight protein marker can be applied to a piece of paper which can be placed on the top slab of the gel next to one end of the IPG strip. One fragment of the gel-less end of the IPG strip can also be cut short.

13. To melt, the agarose sealing solution in 2.0-mL tubes heats up to 50–60 °C in the heating block prior to use.
14. With these settings, the runs should be completed in about 6 h. For longer runs, about 18 h, set the power supply to 80 V constant, and the current and wattage should be set not to exceed 500 mA and 100 W.

References

1. O'Farrel PH (1975) High resolution two-dimensional electrophoresis of proteins. J Biol Chem 250:4007–4021. https://doi.org/10.1016/S0021-9258(19)41496-8
2. Klose J (1975) Protein mapping by combined isoelectric focusing and electrophoresis of mouse tissues. A novel approach to testing for induced point mutations in mammals. Humangenetik 26:231–243. https://doi.org/10.1007/bf00281458
3. Bjellqvist B, Ek K, Righetti PG, Gianazza E, Görg A, Westermeier R, Postel W (1982) Isoelectric focusing in immobilized pH gradients: principle, methodology and some applications. J Biochem Biophys Methods 6:317–339. https://doi.org/10.1016/0165-022X(82)90013-6
4. Kopeć P, Hornyák M, Pastuszak J, Szczerba A, Rapacz M, Waga J, Płażek A (2021) Changes in the flower and leaf proteome of common buckwheat (*Fagopyrum esculentum* Moench) under high temperature. Int J Mol Sci 22:2678. https://doi.org/10.3390/ijms22052678
5. Neuhoff V, Arold N, Taube D, Ehrhardt W (1988) Improved staining of proteins in polyacrylamide gels including isoelectric focusing gels with clear background at nanogram sensitivity using Coomassie brilliant blue G-250 and R-250. Electrophoresis 9:255–262. https://doi.org/10.1002/elps.1150090603

Chapter 13

Measurement of the Kinetics of Chlorophyll *a* Fluorescence by an LED-Light Source Fluorimeter, Handy PEA

Marta Hornyák and Agnieszka Płażek

Abstract

The chlorophyll *a* fluorescence measurement method is used to determine the efficiency of the photosynthetic apparatus and to assess the physiological state of photosynthetic organisms. The measurement is simple, fast, and noninvasive. It is a precise tool to study photosynthesis response under stress conditions or to assess the impact of specific environmental factors on plants. Here we describe the usage of this method in environmental-controlled plant production systems differing in temperature or light source on the growth and development of common buckwheat.

Key words Chlorophyll fluorescence, Noninvasive method, Photosynthesis, Photosystem II, Stress factor

1 Introduction

The analyses of chlorophyll fluorescence have become a powerful and widely used technique used to assess the physiological state of plant physiology and ecophysiology studies. The light absorbed by chlorophyll particles can undergo three fates: [A] photosynthesis, [B] energy dissipated as heat, or [C] energy reemitted as light in the fluorescence process (Fig. 1).

$$\textit{Absorbed energy in PAR range} = \textit{Photochemistry} + \textit{Heat} + \textit{Fluorescence}$$

These three processes are interdependent, so if one parameter increases, another decreases. Therefore, by measuring the yield of chlorophyll fluorescence, the information about the photosynthetic efficiency or energy dissipation can be collected [2]. The phenomenon of chlorophyll fluorescence induction takes place when after darkening the tested photosynthetic organ (e.g., a leaf) for 30 min; will be illuminated with light in the photosynthetically active

Alexander Betekhtin and Artur Pinski (eds.), *Buckwheat: Methods and Protocols*, Methods in Molecular Biology, vol. 2791, https://doi.org/10.1007/978-1-0716-3794-4_13,

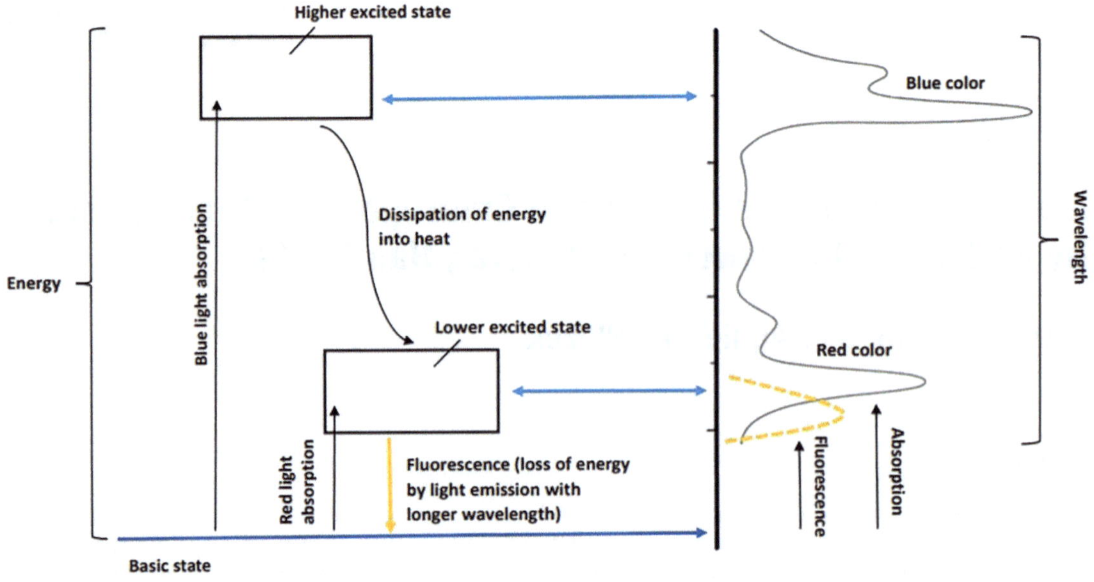

Fig. 1 Absorption and emission of light by chlorophyll [1] diagram with modifications]

radiation (PAR) range. Even if the total amount of chlorophyll fluorescence is very little of the total light absorbed, the measurement is significant. The fluorescence of chlorophyll *a* is a reemission of energy absorbed by chlorophyll *a* molecules. In the leaves, the chlorophyll *a* content is three times higher than that of chlorophyll *b*. Therefore, the fluorescence from the fully developed leaf originated mostly from chlorophyll *a* molecules located in the photosystem II (PSII) [3].

Using this method, we are able to analyze the effect of thermal stress or light source on the photosynthetic efficiency of common buckwheat plants grown in different environment-controlled plant production systems. The measurements should be done minimum on ten plants per treatment [4, 5].

2 Materials

The measurements of the studied parameters should be performed in the fully developed young leaves, for example, the third in the order from the top inflorescences of 8-week-old plants at the full flowering stage. The LED-light source fluorimeter Handy PEA (Hansatech Ltd., King's Lynn, UK) is usually used to perform the measurements of the kinetics of chlorophyll *a* fluorescence.

3 Methods

Prior to the measurements, the LED-light source of a fluorimeter is calibrated using a serial quantum sensor (SQS) light meter. Excitation irradiance intensity is 3000 [μmol m^{-2} s^{-1}] with a peak at 650 nm. Measurements are done after 30 min of the leaf adaptation to darkness by using leaf clips with a 4-mm-diameter hole which is covered using a shutter plate (Fig. 2). Changes in fluorescence are recorded during irradiation between 10 μs and 1 s. Over the initial 2 ms, the data are collected every 10 μs with 12-bit resolution. After this term, the frequency of measurements is reduced automatically. F_v/F_m (quantum yield of PSII) is calculated as $(F_m - F_0)/F_m$, where F_0 and F_m represent the minimal and maximal chlorophyll fluorescence, respectively [6]. These measurements are used to calculate the following parameters based on the theory of energy flow in PSII and the JIP test [7, 8]: ABS/CS_m, energy absorption by antennas; DI_o/CS_m, energy dissipation from PSII; ET_o/CS_m, the energy used for electron transport; PI, performance index of PSII; RC/CS_m, number of active reaction centers; TR_o/CS_m, excitation energy trapped in PSII; ψ_{Ro}, probability, at time 0, that a trapped exciton moves an electron into the electron transport chain beyond QA^-; δ_{Ro}, efficiency with which an electron can move from the reduced intersystem of electron acceptors to the PSI end electron acceptors; and φ_{Ro}, quantum yield of electron transport from QA^- to the PSI end electron acceptors.

To perform the measurements, please carry out all steps as followed:

1. Turn on the fluorimeter Handy PEA+ by button Power *On/Off*.

Fig. 2 The leaf adapts to darkness by using leaf clips with a 4-mm-diameter hole, which is covered using a shutter plate

2. Remove the shutter plate from the diameter hole on the leaf clip.
3. On the visible menu screen, press the commend Measure by *OK* button, or use the remote trigger button placed on Handy PEA Sensor Head.
4. Save the measurement by pressing the *OK* button again.
5. The alphanumeric keypad is used to input the ID of the file. Pressing the key once will input the corresponding numeric value, while further key presses in rapid succession will scroll through the letters assigned to that particular key in the same manner as a cellular telephone keypad.
6. The list of the measurements is available under the Files option on the Main Menu list.
7. The button *Reset* enables you to return to the Main Menu from anywhere in the menu structure.

To transfer the data to your PC, please carry out all steps as follows:

1. Information on the appropriate software can be found on the producer's website: https://www.hansatech-instruments.com/download-category/software/ [9].
2. Using the attached serial cable, connect the Handy PEA to an available serial port (or USB port by using an adapter).
3. Once the data file has been saved and the Main Menu is on, move the screen cursor down to the PC Mode option and press the *OK* key.
4. If the connection has been made, run the PEA Plus software on your PC.
5. Open the Handy PEA drop-down menu again, and select the Download Recordings option.
6. Select the required file range in the displayed dialog, and click OK to download the recording just performed.
7. If the download has been successful, the Summary Screen will be displaying a single record. It is now possible to view this record in the various analysis screens of PEA Plus software.

4 Notes

1. In order to minimalize waiting time before measurement, a number of leaves may be dark adapted simultaneously using several leaf clips.
2. Choose healthy, fully developed leaves for each measurement.
3. Apply the leaf clip on the leaf blade, avoiding leaf veining.

4. Adapt the leaf to darkness by using leaf clips for 30 min before measurement. Try to keep each leaf in darkness for the same amount of time.
5. Before each measurement, be sure to remove the shutter plate from the diameter hole on the leaf clip.
6. Make sure that each record is saved.
7. If any record is not clear, repeat the measurement.
8. Tips to help you interpret your results [3, 10]

CS_m: the cross-sectional area of the leaf blade

PI (ABS): performance index (on an absorption basis), an indicator informing about the efficiency of changes in PSII

PI (ABS) = forces proceeding photochemical transformations/forces inhibiting these transformations

Therefore, a high value of this indicator is desirable.

ABS/CS_m: photon flux absorption by chlorophyll molecules of both active and inactive RC of PSII. A decrease in these values indicates a decrease in energy absorption efficiency.

DI_o/CS_m: indicates the amount of energy dissipation (energy lost as heat).

$$DIo/CSm = ABS/CSm - TRo/CSm$$

An increase in energy dissipation under stress conditions demonstrates an increase in energy loss and a decrease in the energy necessary for photochemical transformations.

ET_o/CS_m: energy stream (e) transported out of QA. Under stress conditions, a decrease in the value of this indicator is typical.

RC/CS_m: density of active reaction centers (RCs) of PSII. Under stress conditions, a decrease in the value of this parameter is observed.

TR_o/CS_m: flux of captured energy due to the reduction of Q_A to Q_A^-. A decrease may indicate that active RCs are converting to inactive RCs.

Acknowledgments

This work was supported by the National Centre of Science in Poland (no. 2017/25/B/NZ9/00148).

References

1. Filek W, Kościelniak J, Dubert F, Rapacz M, Skrudlik G (2007) Ćwiczenia z fizjologii roślin z podstawami biochemii. Wydawnictwo Uniwersytetu Rolniczego w Krakowie
2. Maxwell K, Johnson GN (2000) Chlorophyll fluorescence—a practical guide. J Exp Bot 51(345):659–668. https://doi.org/10.1093/jexbot/51.345.659
3. Kalaji MH, Łoboda T (2009) Chlorophyll fluorescence to in plants' physiological state researches. Warsaw University of Life Sciences—SGGW, Warsaw

4. Hornyák M, Płażek A, Kopeć P, Dziurka M, Pastuszak J, Szczerba A, Hura T (2020) Photosynthetic activity of common buckwheat (*Fagopyrum esculentum Moench*) exposed to thermal stress. Photosynthetica 58(1):45–53. https://doi.org/10.32615/ps.2019.140
5. Hornyák M, Dziurka M, Kula-Maximenko M, Pastuszak J, Szczerba A, Szklarczyk M, Płażek A (2022) Photosynthetic efficiency, growth and secondary metabolism of common buckwheat (*Fagopyrum esculentum* Moench) in different controlled-environment production systems. Sci Rep 12(1):257. https://doi.org/10.1038/s41598-021-04134-6
6. van Kooten O, Snel JFH (1990) The use of chlorophyll fluorescence nomenclature in plant stress physiology. Photosynth Res 25:147–150. https://doi.org/10.1007/BF00033156
7. Lazár D (1999) Chlorophyll *a* fluorescence induction. BBA. Bioenergetics 1412:1–28. https://doi.org/10.1016/s0005-2728(99)00047-x
8. Strasser RJ, Srivastava A, Tsimilli-Michael M (2000) The fluorescence transient as a tool to characterize and screen photosynthetic samples. In: Yunus M, Pathre U, Mohanty P (eds) Probing photosynthesis: mechanism, regulation and adaptation. CRC Press, New York, pp 445–483
9. Hansatech Instruments website. https://www.hansatech-instruments.com/download-category/software/
10. Kalaji HM, Schansker G, Brestic M, Bussotti F, Calatayud A, Ferroni L et al (2017) Frequently asked questions about chlorophyll fluorescence, the sequel. Photosynth 132:13–66. https://doi.org/10.1007/s11120-016-0318-y

Chapter 14

Measurements of Leaf Gas-Exchange Parameters Using Portable CIRAS-3 Infrared Gas Analyzer, with a Parkinson Leaf Chamber (PLC6)

Marta Hornyák, Maciej Grzesiak, and Agnieszka Płażek

Abstract

Ranges of portable systems to measure leaf gas-exchange parameters are available. They allow real-time measurements of the photosynthesis rate (A), transpiration rate (E), stomatal conductance (g_s), and intercellular CO_2 concentration (C_i). Photosynthetic CO_2 uptake is one of the most frequently studied plant physiological processes. The measurement is precise, simple, and noninvasive to perform *in vivo*. We describe the use of this method in environmental-controlled plant production systems at different temperatures on the growth and development of common buckwheat.

Key words Leaf gas exchange, Noninvasive method, Photosynthesis, Stress response

1 Introduction

The gas analyzer is an assembly of instruments for detecting the amount of CO_2 assimilated by the leaf, transpiration H_2O, and temperature of the air and leaf. These data are used to calculate the net photosynthesis rate, stomatal conductance, transpiration rate, and intercellular CO_2 concentration [1]. Important factors in the biomass production of plants include net photosynthetic CO_2 assimilation per unit of area and time (A_{net}), leaf area, and distribution. Therefore, the leaf gas-exchange parameters are commonly used in plant physiology to better understand interspecific differences in photosynthesis and transpiration and to define the rapid response to changes in environmental factors such as light, humidity, or temperature and response to various abiotic and biotic stresses [2].

Using this method, we can analyze the effect of thermal stress on photosynthetic efficiency and growth of common buckwheat

Alexander Betekhtin and Artur Pinski (eds.), *Buckwheat: Methods and Protocols*, Methods in Molecular Biology, vol. 2791,
https://doi.org/10.1007/978-1-0716-3794-4_14,

plants grown in different controlled and environmental plant production systems. The measurements should be done minimum on ten plants per treatment [3].

2 Materials

The measurements of the studied parameters should be performed on the fully developed young leaves, for example, the third in the order from the top inflorescences of 8-week-old plants at the full flowering stage. The CIRAS-3 infrared gas analyzer (PP Systems, MA, USA), with a Parkinson leaf chamber (PLC6), is used to perform the measurements.

3 Methods

Before the measurement, the irradiation system consisting of halogen lamps was attached. The airflow rate with a constant CO_2 concentration of 400 μmol mol^{-1} through the assimilation chamber was 300 cm^3 min^{-1}. The temperature of the leaves was 22 °C, the air humidity was 40%, and the irradiance was 800 μmol photon m^{-2} s^{-1}. To ensure the same conditions, and the circadian rhythm, all measurements were carried out between 9 and 11 am. All measurements were taken in eight replicates for each accession and treatment. The following parameters were determined and calculated: the rate of net photosynthesis (A) and transpiration (E), stomatal conductance (g_s), and internal CO_2 concentration (C_i). Apparent carboxylation efficiency (A/C_i) was calculated according to Niu et al. [4] and instantaneous (*WUE*, A/E) and intrinsic water use efficiencies (WUE_i, A/g_s) according to Medrano et al. [5].

To perform the measurements, please carry out all steps as follows:

1. Press and release the *ON/OFF* button; it should illuminate blue.
2. Wait for the Welcome display to appear on the console screen.
3. Press either Continue (*F1*) or Settings (*F2*) – "Warm-up (temp) C..." appears in the status bar at the bottom of the screen.
4. Warm-up time to the required 55 °C IRGA can take a few minutes; wait until done.
5. Once the IRGA achieves its target temperature of 55 °C, it will then perform both a Zero and Diff Bal cycle.
6. To enter the Numeric View screen, press Continue (*F1*).

Fig. 1 Placing the leaf blade in the cuvette of the gas analyzer

7. Now that you can observe live readings, you can check the system for stability and leaks. This is usually done with the leaf chamber empty. Press Controls (*F3*), and enter the CO_2 concentration that you might be working with, and use the default H_2O Reference control option "Fixed % of reference" and 100%. Accept the change to return to the Numeric View screen (*see* **Notes 1–6**).
8. You can place the leaf in the cuvette (Fig. 1). Try approaching the leaf to avoid accidental injury of its blade. You can perform the measurement (*see* **Notes 7–10**).
9. If your goal is to collect a large quantity of relatively short-duration measurements from a large sample group of plants/leaves, you will probably use the Manual Recording option as we did, or you can use mini-USB for data transfer to an external PC.
10. For a detailed manual, please check the information sheet for the portable CIRAS-3 infrared gas analyzer (PP Systems, MA, USA) at the producer's website [6].

4 Notes

1. Be sure that the chemicals (molecular sieve, CO_2 cartridges) are fresh.
2. Verify if the leaf chamber pads are in good condition.
3. Check if there is sufficient battery power for field measurement.
4. Check all necessary electrical connections of cuvettes and other components.
5. Keep the CIRAS-3 on its power supply during warm-up to preserve the battery.
6. While using a CO_2 cartridge, check if that one is in place with enough capacity for all your measurements.

7. While doing the measurement, try to orient the cuvette to the leaf and do not twist the petiole.
8. Use a tripod if possible; you will have both hands free and avoid changes to the natural leaf position.
9. Consider that some physiological variables, especially A and g_s, may continue changing as the leaf acclimates and approaches a stable state.
10. For more information and tips, check the most recent manual for portable CIRAS-3 infrared gas analyzer on the producer's website [6].

5 Interpretation of the Results

1. Based on the net photosynthesis rate and transpiration rate, a coefficient WUE (water use efficiency) can be calculated. Under controlled and optimal growth conditions, WUE is specific to each plant species, but under stress conditions, the higher the value of this parameter, the better the water use and the higher the plant resistance to stress.
2. Under some stress conditions, in particular, under osmotic stress (e.g., drought, salinity), the stomata are close to reducing the transpiration rate [7, 8]. This would be expected to reduce the efficiency of CO_2 assimilation from the air. However, there is often no significant decrease in the net photosynthesis rate associated with stomatal closure. This fact can be interpreted as the use of carbon dioxide from photorespiration. This effect can be also noticed in C3 plants and explained by the activation of phosphoenolpyruvate carboxylase, which is involved in the first carboxylation in the dark phase of photosynthesis in C4 plants. In addition, with the closing of the stomata during active carboxylation, the CO_2 content inside the leaf decreases, which in turn increases the gradient of its concentration in the intercellular spaces and the atmosphere resulting in faster diffusion of CO_2 into the leaf.

Acknowledgments

This work was supported by the National Centre of Science in Poland (No. 2017/25/B/NZ9/00148).

References

1. Pandey R, Paul V, Singh MP (2017) Measurement of photosynthesis using an infrared gas analyzer. Manual of ICAR Sponsored Training Programme for Technical Staff of ICAR Institutes on Physiological Techniques to Analyze the Impact of Climate Change on Crop Plants.

IIFSR, p 19. https://doi.org/10.13140/RG.2.2.29626.90561

2. Duursma RA (2015) Plantecophys-an R package for analyzing and modeling leaf gas exchange data. PLoS One 10(11):e0143346. https://doi.org/10.1371/journal.pone.0143346
3. Hornyák M, Płażek A, Kopeć P, Dziurka M, Pastuszak J, Szczerba A, Hura T (2020) Photosynthetic activity of common buckwheat (*Fagopyrum esculentum* Moench) exposed to thermal stress. Photosynthetica 58(1):45–53. https://doi.org/10.32615/ps.2019.140
4. Niu SL, Li LH, Jiang GM et al (2004) Gas exchange and chlorophyll fluorescence response to simulated rainfall in *Hedysarum fruticosum* var. *mongolicum*. Photosynthetica 42:1–6. https://doi.org/10.1023/B:PHOT.0000040562.34339.b3
5. Medrano H, Tomás M, Martorell S et al (2015) From leaf to whole-plant water use efficiency (WUE) in complex canopies: limitations of leaf WUE as a selection target. Crop J 3:220–228. https://doi.org/10.1016/j.cj.2015.04.002
6. PP Systems website. https://www.ppsystems.com
7. Pastuszak J, Dziurka M, Hornyák M, Szczerba A, Kopeć P, Płażek A (2022) Physiological and biochemical parameters of salinity resistance of three durum wheat genotypes. Int J Mol Sci 23(15):8397. https://doi.org/10.3390/ijms23158397
8. Płażek A, Dubert F, Kościelniak J, Tatrzańska M, Maciejewski M, Gondek K, Żurek G (2014) Tolerance of *Miscanthus* × *giganteus* to salinity depends on initial weight of rhizomes as well as high accumulation of potassium and proline in leaves. Ind Crop Prod 52:278–285. https://doi.org/10.1016/j.indcrop.2013.10.041

Chapter 15

Measurement of the Light Intensity and Spectrum Influence on Plant Growth and Secondary Metabolites of Common Buckwheat

Monika Kula-Maximenko, Marta Hornyák, and Agnieszka Płażek

Abstract

Light is one of the main signals detected by plants that influence plant growth, development, and function. The light features that influence plants are the photoperiod, light intensity, and spectral composition. Manipulating light intensity and spectrum to obtain better plant growth and quality has become a popular research object in recent years. Here we describe the usage of the spectrometer Lighting Passport Pro to determine the impact of light intensity and share of individual waves in its spectrum in environment-controlled plant production systems on the growth, development, and soluble carbohydrate and phenolic synthesis of common buckwheat.

Key words Noninvasive method, LEDs, Light spectrum, Photosynthesis, Wavelengths

1 Introduction

Light as a primary energy source is one of the most important environmental factors for plant development [1]. The light intensity and spectrum are crucial for plants' growth, morphogenesis, and multiple physiological functions [1, 2, 3]. The study about light intensified due to the progressive development of computer-controlled systems for artificial illumination and measuring photosynthetic activity undertaken more and more. Insights into the molecular mechanisms of photo-regulation of physiological functions in plants mediated by photoreceptors and other pigments require special monitoring of light conditions [4].

Measurement of the light intensity and spectrum is used to analyze the effect of the light source on the photosynthetic efficiency of common buckwheat plants grown in different environment-controlled plant production systems (Fig. 1). We

Alexander Betekhtin and Artur Pinski (eds.), *Buckwheat: Methods and Protocols*, Methods in Molecular Biology, vol. 2791,
https://doi.org/10.1007/978-1-0716-3794-4_15,

Fig. 1 The measurement of light intensity and spectrum by the spectrometer Lighting Passport Pro

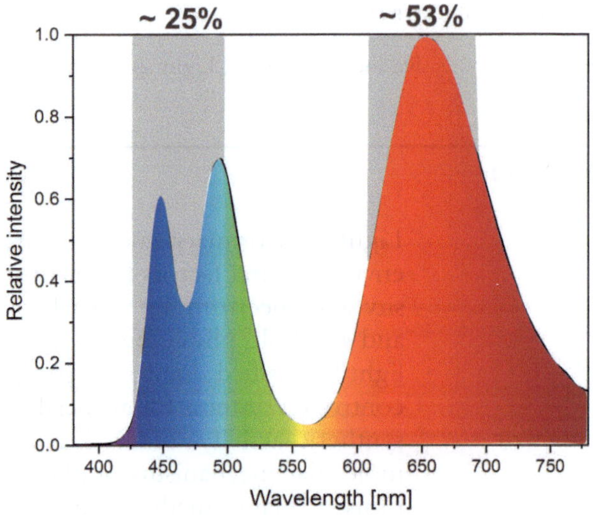

Fig. 2 An example of a spectrum of LED light with a marked range of light relevant to obtaining a higher phenolic acid content in buckwheat

were analyzing the effect of solo light emitting diodes (LEDs) in phytotronic chamber and solar radiation supplemented with high-pressure sodium (HPS) lamps in a greenhouse as a control [5] (Fig. 2).

2 Materials

The measurements of light intensity and spectrum were performed on the spectrometer Lighting Passport Pro (Asensetek, Taiwan) with Spectrum Genius Cloud (Taiwan) software for both plants [5, 6] and algae [7].

3 Methods

Before the light measurement, we need to download Spectrum Genius Agricultural Lighting (SGAL App) application on our smartphone or tablet. First, you should turn on the power of the Lighting Passport, which power button is on the left side of the device. Then, the blue diode on the back of the device will turn on. The color of the diode on the back of the device informs about the battery charge level: blue, fully charged; green, half discharged battery; and red, discharged battery, connect the charger. To connect a smartphone or tablet to the application, you should turn on Bluetooth. In the SGAL App, tap "Bluetooth Pairing" in the settings, and select the device number that has been found to connect it. If your device is not found, please press the "Rescan" button, and wait until your device number is displayed. Before measuring, you can set the following parameters: ID, image, temperature, and humidity.

3.1 Measurements

To perform the measurements, please carry out all steps as follows:

1. Before measuring, you must open the ND filter cover. The white ND filter is located on top of the device and is covered by a sliding black slide.
2. You place the device at leaf height. The ND filter must be under a test light source and cannot be obscured by leaves.
3. Go to the application, after the device is in the right place.
4. You will see six windows on the application's main page. Please select one of them depending on what measurement you want to perform: Diary, if you want a continuous record of growth and check the changes in the object; Single, measurement and information only for a single light source; and Multiple, measurement and information for a multiple light source. If you choose single or multiple, the steps are similar.
5. When you tap the "Single" or "Multiple" button, you can choose your default PAR reference or change the default reference before the measurement by selecting the Ref button. Select a reference and press the "Next" button.
6. If you do not want a reference spectrum, press the "Next" button.

7. Tap the "Measure" circular button to start your measurement.
8. After that, you will see the PAR action light spectrum results and PPFD in each wavelength.
9. You can select which on the specific waveband spectrum must be shown by the five buttons: blue (B), green (G), red (R), ultraviolet (UV), or far-red (FR).
10. On the next page (after moving the screen to the left), all measured parameters will be displayed: PPFD, YPFD, efficiency for the entire spectrum, and individual wavebands. In addition, we will get the R/FR, R/B, DLI, and illuminance [lux] values.
11. Press the "Next" button to save the result after the measurement.
12. You can photograph the plant and save detailed measurement information.
13. Save the measurement by pressing the "Save" button.
14. You repeat the measurement for the light source in several places to have an average for the whole plant.

3.2 Data Transfer

To transfer the data to your PC, please carry out all steps as follows:

1. All performed results are saved in the "Record(s)" window.
2. The results can be e-mailed directly after each measurement or after a series of measurements by entering the "Record(s)" window.
3. Measurements saved in "Record(s)" can be opened, analyzed, and shared by e-mail.
4. To send the measurement results by e-mail, select a single measurement and press the "Share" button.
5. The data will be sent as a ".csv" file.

4 Notes

Measurement capability.

Photosynthetic photon flux density (PPFD) and yield photon flux density (YPFD) [μmol/m^{2}s]: for the whole spectrum (400 nm ~ 700 nm), far-red (701 nm ~ 780 nm), red (600 nm ~ 700 nm), green (500 nm ~ 599 nm), blue (400 nm ~ 499 nm), UV (380 nm ~ 399 nm); weighted spectrum; efficiency [%]; the ratio of red/blue light; the ratio of red/far-red light; daily light integral (DLI) [mol/m^2]; illuminance [lux]; peak wavelength (λp) [nm]; dominant wavelength (λD) [nm]; correlated color temperature (CCT) [K]; color rendering index (CRI (Ra)); CIE 1931/CIE 1976.

Acknowledgments

This work was supported by the National Centre of Science in Poland (No. 2017/25/B/NZ9/00148).

References

1. Fukuda N, Fujita M, Ohta Y et al (2008) Directional blue light irradiation triggers epidermal cell elongation of abaxial side resulting in inhibition of leaf epinasty in geranium under red light condition. Sci Hortic 115(2):176–182. https://doi.org/10.1016/j.scienta.2007.08.006
2. Li Q, Kubota C (2009) Effects of supplemental light quality on growth and phytochemicals of baby leaf lettuce. Environ Exp Bot 67(1):59–64. https://doi.org/10.1002/jsfa.6173
3. Fan XX, Xu ZG, Liu XY et al (2013) Effects of light intensity on the growth and leaf development of young tomato plants grown under a combination of red and blue light. Sci Hortic 153:50–55. https://doi.org/10.1016/j.scienta.2013.01.017
4. Ptushenko OS, Ptushenko VV, Solovchenko AE (2020) Spectrum of light as a determinant of plant functioning: a historical perspective. Life 10(3):25. https://doi.org/10.3390/life10030025
5. Hornyák M, Dziurka M, Kula-Maximenko M et al (2022) Photosynthetic efficiency, growth, and secondary metabolism of common buckwheat (Fagopyrum esculentum Moench) in different controlled-environment production systems. Sci Rep 12(1):257. https://doi.org/10.1038/s41598-021-04134-6
6. Kula-Maximenko M, Niewiadomska E, Maksymowicz A et al (2021) Insight into details of the photosynthetic light reactions and selected metabolic changes in tomato seedlings growing under various light spectra. Int J Mol Sci 22:11517. https://doi.org/10.3390/ijms222111517
7. Kula-Maximenko M, Zieliński KJ, Ślesak I (2021) The role of selected wavelengths of light in the activity of photosystem II in *Gloeobacter violaceus*. Int J Mol Sci 22:4021. https://doi.org/10.3390/ijms22084021

INDEX

A

Agarose 46–48, 50, 51, 54, 55, 73, 77–79, 115, 117, 119
Aniline blue staining for callose 98
Antibody 15–18, 20, 21, 58–60, 63, 65, 69, 71, 72, 74, 77–79
Arabinogalactan proteins (AGPs) 57, 59, 67, 71

B

Buckwheat 2–13, 35–42, 45–56, 81, 85, 86, 89–91, 93, 95, 97–102, 105, 108, 113, 114, 122, 127, 133–136

C

Callus 2–13, 16, 18, 21, 24, 35, 36, 40, 50, 51, 54, 55, 90, 95
Cell wall 36, 45, 57, 58, 69, 71–80, 85, 98, 103, 104
Chlorophyll fluorescence 121–123
Chromosomes 81, 83, 84, 86
Colloidal Coomassie staining 114, 117, 118
Confocal microscopy 23–33, 78

D

DNA methylation 16, 20, 23, 24
Driselase 47, 49

E

Embryo 48, 51, 55, 89–92, 94–99
Embryogenesis 96, 99, 100
Embryo rescue 89, 90, 95, 96
Enzyme cocktail 82, 83, 86
Epigenetics 15–21, 23–33

F

Fagopyrum 15–21, 23–33
Fagopyrum esculentum 2, 6, 7, 11, 16, 18, 24, 51, 57–69, 71–80, 82–86, 89, 95, 97, 99, 100, 114
Female gametophyte 99, 101–104
Ferric chloride 36

H

Hairy roots 2–13
Histone acetylation 24
Hypocotyls 6, 46, 48, 49, 53, 96

I

Immunostaining 15–18, 20, 23, 60, 63, 65, 69, 71, 74, 77, 79
In vitro cultures 89
Isoelectric focusing (IEF) 113, 114, 116, 118

L

Leaf gas exchange 127–130
Light emitting diodes (LEDs) 134
Light spectrum 78, 136

M

Micronuclei 81, 83–86
Morphogenic callus (MC) 7, 16, 18, 36, 39, 46, 50, 51, 54–56

O

Osmium tetroxide (OsO_4) 35–41
Ovary 91, 92, 95, 96, 99, 100
Ovule 91, 92, 94–96, 98–100, 105

P

Paraffin method 100–104
Pectins 57, 71
Phenol extraction 107
Phenolic 35, 36, 39–42, 54, 55, 71, 107, 109, 110, 134
Photosynthesis 121, 127, 128, 130
Photosystem II (PSII) 47, 49, 50, 53, 54, 122, 123, 125
Phytosulfokine (PSK) 46, 48, 50, 52, 54, 55, 73, 75
Plant cells 2, 45, 57, 58, 65, 71
Plantlet 7, 92–94, 96
Plant regeneration 11, 45, 50, 51, 54, 55

Alexander Betekhtin and Artur Pinski (eds.), *Buckwheat: Methods and Protocols*, Methods in Molecular Biology, vol. 2791, https://doi.org/10.1007/978-1-0716-3794-4,

Pollen stainability by acetocarmine and Alexander dye 104
Pollen tubes 97, 98, 100, 104
Protein precipitation 109
Protein purification 109
Protein solubilization 109
Proteome 108, 113–119
Proteome analysis 107, 113
Protoplasts 45–56, 72–73, 75–80, 104

R

R Studio 23–33

S

Seedling 6, 9, 61, 83, 91, 92, 94–96
Semi-thin sections 36, 39, 41
Sodium dodecyl sulfate-polyacrylamide gel electrophoresis (SDS-PAGE) 113, 114, 116
Stress response 127

T

Toluidine blue O (TB) 35–41, 60
Two-dimensional gel electrophoresis (2-DE) 107, 113, 114

W

Wavelengths 65, 78, 136

MIX
Papier aus verantwortungsvollen Quellen
Paper from responsible sources
FSC® C105338

If you have any concerns about our products,
you can contact us on
ProductSafety@springernature.com

In case Publisher is established outside the EU,
the EU authorized representative is:
Springer Nature Customer Service Center GmbH
Europaplatz 3, 69115 Heidelberg, Germany

Printed by Libri Plureos GmbH
in Hamburg, Germany